Analysing Cycles

Razorbill Press

The Press is named after a seabird, the Razorbill (*Alca torda*). Locally in Newfoundland it's also known as the "Tinker" (thinker), because of the way it holds its head, appearing lost in thought, looking to the distance somewhat above the horizon. Razorbill and Tinker have a pleasant humour about them, and Tinker also carries the meaning of tinkering, experimentation, investigation, which fits nicely with the curiosity that founds science.

"Tinker" therefore is an imprint of Razorbill Press for books that relate to science and explorations.

For news, updates, supplementary materials, other titles, and notices, visit:

Razorbill Press
http://www.razorbillpress.com/

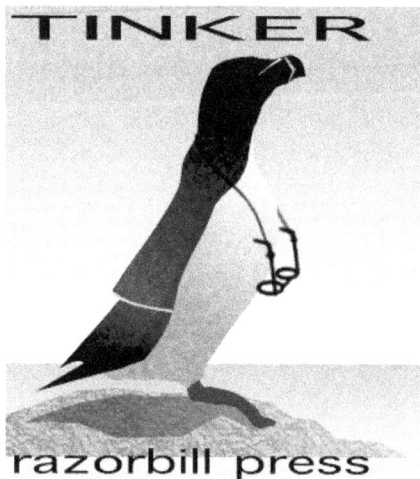

ANALYSING CYCLES

IN BIOLOGY & MEDICINE

A PRACTICAL INTRODUCTION TO
CIRCULAR VARIABLES & PERIODIC REGRESSION

SECOND EDITION

K.N.I. Bell, *B.Sc., M.Sc., Ph.D.*

Department of Ichthyology and Fisheries Science, Rhodes University

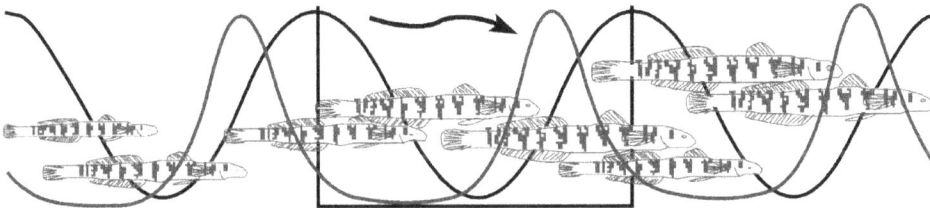

Recruitment dynamics of Sicydium punctatum became understandable through periodic regression. The first, sinusoidal, curve is age of migrating fish; the second derives from the first and is a good match for numbers migrating.

(Can.J.Fish.Aquat.Sci. 52:1535-1545 & 54:1668–1681)

Paperback edition: ISBN-13:   978-0-9736209-2-4

                       ISBN-10:   0-9736209-2-7

Cite this book as:
Bell, K.N.I., 2008. Analysing cycles in biology & medicine—a practical introduction to circular variables & periodic regression. Second edition. Razorbill Press, St. John's, Newfoundland, Canada A1C 3T2.

Previous edition information: first edition published 2004 under title: Introduction to Circular Variables & Periodic Regression in Biology (electronic book in PDF format), ISBN 0-9736209-0-0.

version b (AC2:4.1.Lf)

About the author

The author's degrees are from Dalhousie University and Memorial University of Newfoundland. He has given seminars, courses, and workshops—in biology, introductory statistics, data management, and science ethics—at Dalhousie, Memorial, Rhodes, and elsewhere. His core interests are anadromous gobies, recruitment processes, early life histories, and ultimately conservation. He argues for openness of science, for science ethics in government decision-making, and for accountability in policy and its implementation. He has consistently pointed out the problems that result from secrecy, as argued by the venerable C.P. Snow (1962) in *Science and Government*. He has reviewed for many journals, reported to the House of Commons Standing Committee on Fisheries and Oceans, and, for COSEWIC (Committee on the Status of Endangered Wildlife in Canada), authored the 1998 Status Report on Atlantic Cod. That Report resulted in Canada's first at-risk designation of a commercial marine fish. But that is another story.

From the author: history and thanks

Cycles are the essence of life. I am a field ecologist who ran into periodic data that his colleagues couldn't advise on. I therefore improvised, re-invented, and later linked with established methods. It became increasingly clear that biologists need to know how to analyse cycles.

Going back to where it began, I had seen wind directions improperly summarized by the arithmetic mean; that was clearly wrong because the arithmetic mean of a few northerly winds can give a southerly result. I commented on it to Dave Schneider, who replied that the proper way to average directions was to separately average the sines and cosines. (Thank you, Dave, for your broad interests.)

This book thus began as notes to myself and a letter to Mr. Marc Blanc of the weather service of Dominica, who merits thanks for his dedication to preserving historic weather data. The notes grew to include periodic regression in response to my own needs, and then in response to questions from others.

I initially improvised periodic regression by iteratively fitting a cosine function. I worried, however, that I might be underestimating DF because I had fitted the lag before doing the final regression analysis. DF and significance are tightly bound, and any possible exaggeration of significance is a concern. While most colleagues responded that I was merely fitting a function (and indeed one eminent text makes the same mistake), W.G. Warren—whose joy in statistical methods is positively infectious—said yes, of course you need that penalty DF, and by the way why are you using that long iterative procedure? He scribbled an alternative on the whiteboard. At first glance it seemed alien, and I said so, adding that perhaps a while later it might become clear, which it did, almost effortlessly. Thank you, Bill, for your insightful and patient help in many discussions.

Thanks are due many more. A.J. (Tony) Booth introduced me to the art of macros and he and Horst Kaiser read key sections several times. I thank Ted Miller, Alan Whittick, Phil Heemstra, Emmanuel Kaunda, Chris McCartney, Sloans Chimatiro, Olaf Weyl, Ivan Mateo, Daniel Pauly, Dominique Ponton, Ntobeko 'Bucs' Bacela, 'Uncle' Ben Ngatunga, students at an impromptu workshop at U. Philippines Diliman, Peter Scott, Peter Earle, Alan Richards, Bill Montevecchi, and others whose comments, needs, questions, and advice have helped push this book along. The author photo was taken by Ian Jones; Antony Bell, Peggy Stewart and Aaron Shepard helpfully commented on presentation; J.R. Bell proofread hundreds of pages. I thank Chester I. Bliss for his manual of 1958 and Edward Batschelet for his book of 1981. More generally, my philosophy benefited from Ian McLaren, Pat Lane, Bob Rosen, Bob Rodger, Sifford Pearre, Pierre Pepin, Jean-Marie Sevigny, Ron O'Dor, Bob Boutilier, Dick McBride, Roger Doyle, Dave Patriquin, Iain Suthers, Nils Hagen, Ron Smith, and others at Dalhousie. Personal thanks are to my family: Jane Robin, Lieda, Peter, Charlotte, Antony, and Christine (McGeoch); Lee and Jake and the "Deorksen Institute", George and Carolyn Mayo, the Bridger family and Bachar Chalati, and many others. Worth savouring is Augustus Mamaril's crisp comment *"of course our students must be smarter than we are; if not, we are not doing our job"*.

Despite help from others, errors remain my responsibility alone. I welcome corrections, comments, suggestions, and postcards.

Kim N.I. Bell, B.Sc.[1], M.Sc[2]. Ph.D.[1] ([1]M.U.N. [2]Dalhousie)

25 Monkstown Road
St. John's, NL
Canada A1C 3T2
kbell@mun.ca or via Razorbill Press
www.ucs.mun.ca/~kbell

CONTENTS

PREFACE

Statistics condenses information to facilitate judgement on function estimates and comparisons. It also provides a philosophy to help communicate across disciplines. To give us an objective framework for judgement, statistics balances the data against what is estimated, limiting the claims that can be made; limiting, if you like, the right to generalise. Therefore, with the help of good statistics, judgement becomes more sound and more useful.

This book exists because, in the biological sciences, the objective analysis of cycles has been too poorly known. We have somehow relegated cycles to the exotic and marginal, and have not exploited their interest and explanatory potential.

In various drafts, this book has been used by a circle of colleagues from South Africa to New Caledonia to Newfoundland, on topics from recruitment, to growth rates, to seasonal allometry, and to estuarine access. The book grew from my notes to myself, and then in response to advice requested by students and colleagues.

The main aim is to introduce periodic regression. We therefore deal mainly with:

- A basic understanding of circular variables like dates and directions (Sec. 3), concepts like trajectories and vectors (Sec. 4), the proper handling of common cycles in analysis (Sec. 5), and conversion from an angle α to the $(\sin\alpha, \cos\alpha)$ pair (Sec. 3.4, 3.5), and (much trickier) vice versa (Sec. 4.3).

- Periodic regression (Sec. 6), analysing variation in a (linear, periodic) Y as a function of variation in a circular-scale X.

Related basic topics are given in support. The variation of a directional Y as a result of a directional or circular X (the "spherical" situation) is beyond the scope of this book.

This book anticipates readers ranging from apprehensive to advanced. The necessary trigonometry is described with application to handling circular variables. Likewise, to help readers ride out difficulties with half-forgotten concepts, there is Sec. 11 (Appendix III: Stats refresher in a rush). There is also a glossary section and index.

Nearly a century ago, Professor Silvanus P. Thompson used a welcoming and generous style in his rigorous yet welcoming book *Calculus Made Easy*. It carried the encouragement "what one fool can do, another can", which he attributes as *'Ancient Simian Proverb'*. Thompson's charming book is still obtainable. I share his wish to demystify and to make accessible a useful analytical toolbox, and I'll shamelessly use repetition, corny jokes, puns, etc. to help do so.

The presentation style is intended to give the reader multiple routes to understanding a concept. There is some restatement to make material accessible without a lot of page-turning, and cross-references to make page-turning more effective. The apprehensive reader should be able to continue past explanations that don't reach him, to explanations that do. (I apologise in advance to those who find too much explanation.)

Graphically-oriented people may find the diagrams sufficient. Conversely, left-brain people might find the equations sufficient and the graphics pointless. A spreadsheet layout helps form a mental image of data structure; sections 6.1, 9.1, 9.2 give example data in spreadsheet form, showing the necessary transformations and how columns represent Xs and Ys.

Conventions vary amongst authors (Sec. 3.2.3). This book favours the 'azimuthal' or compass-like visualisation system (Sec. 3.2), because it taps into familiar examples.

All readers should observe the cautions regarding possible DF errors which even advanced readers may make (and indeed were mistakenly recommended by a major text).

The book is directed toward driving the bus, rather than building it. Modern computing has changed statistics, changed the way we interact with the machinery of it. The machine does the implementation corresponding to our request. We, the researchers and data analysts, must operate at a higher level: the final step in any analysis is judgement.

Statistics needn't be intimidating; there are often several right answers. There is as much variation in style and personality amongst statisticians as in any other group, and many problems can be attacked in more than one way, so few statisticians should be expected to favour the same way.

Finally, to counter those who delight in dismissing stats with that old canard "there are lies, and damn lies, and then there are statistics", tell them it means "there are lies, and also damn lies, but on the other hand there are statistics." Only bad statisticians fall for bad statistics.

MOTIVATION

"Periodic phenomena in biology and climatology occur so widely that we tend either to adapt to them as unavoidable nuisances or are overimpressed by their day to day deviations. We can't see the forest for the trees" — C.I. Bliss, 1958.

Why do cycles matter? Biology is dominated by periodic variation (e.g.: variation according to directions; times in cycles like year, lunar, day, and even tide). Regrettably however, few biologists have literacy in circular statistics. But circular stats are not difficult—'tricky' would be a better word: with a few key concepts and cautions they become easily managed.

The very fact you picked up this book means you had a problem to solve. So, even if *think* you dislike trigonometry, or formulas, or even statistics, try to think of those things not as dislikes, but things you haven't yet acquired a taste for—goodies you haven't yet unwrapped.

Why would we need to analyse cycles? Either (a) to describe (model) in a rigorous way their effect on data; and/or (b) to be able to remove their effect from data that we are exploring for some other non-cyclic question. The same system of analyses will let us do either or both.

Every biologist knows what a seasonal, or daily, or tidal cycle is, but precious few can analyse one. Paradoxically, these cycles, so well known to all of us by experience, are routinely ignored in research plans and analyses. As an example of how poorly-known periodic regression is amongst biologists: when I presented a plot (like Figure 6-5) of a periodic regression at a conference at St. Andrews, a senior and influential biologist retorted "why don't you just put a straight line through that?" He soon got it, but the comment shows why this book is needed.

Typically, cycles are ignored altogether. Sometimes, workers try to exclude cyclic variation in data, e.g. by keeping to one time of day. It is a poor approach, however; it reduces discovery opportunities. With multiple cycles (e.g. daily and tidal) their least common multiples are so large that such approaches require waiting for very long times between sampling (hence wasting opportunities). Even if such an approach is completed with technical success within its narrow terms of reference, its very narrowness renders its conclusions meaningless outside the actual range sampled.

Instead of attempting to fence periodicity out of the data, put it into the analysis! It is more revealing, flexible, and beautiful; it's not difficult, and it's potentially very rewarding. It can simplify the planning of sampling protocols and will likely give better resolution of the variables of interest, better justification for generalisation from results, and, as a bonus, information about the effects of cycles on the variables of interest.

Analysing cycles concerns circular variables and periodic variables. The distinction between these is subtle, but fundamental. Both are often described as cycles, but a variable is circular if it (e.g. time of day) expresses a circular scale, whereas a variable (e.g. temperature) is periodic only where it shows a repeating pattern against a circular variable. Thus, one cycle is an index against which the other, a phenomenon, is measured.

The key to understanding circular variables is they occupy [at least] two dimensions (not one as linear variables do). That means that even simple concepts like averaging require special handling for circular variables, for which the arithmetic mean is clearly nonsensical. Circular variables, and the periodic variables which respond to them, thus require techniques different from those safe for purely linear variables like weight—but it is not difficult, and even a little work can result in profound illumination.

Periodic regression can describe periodic variation in a Y variable according to a circular X variable. Considerable complexity is possible in an analysis, but the sinusoidal curve is the simplest possible periodic pattern. The associated ANOVA can determine the statistical significance of changes in Y as responses to circular X or Xs; linear Xs can be included in the regression as well. Periodic regressions can be used to remove periodic trends from data, in order to make other analyses possible.

Some people—even some spectacularly bright ones—shudder at the very words sine, cosine, and tangent; yet these simple concepts aren't difficult, there aren't very many of them, and they are incredibly useful.

For a start, sine and cosine are, as a pair, simply an alternative description of an angle. In fact, it could be argued they are the most fundamental description of an angle because they are the same no matter what scale—degrees, rads, days-of-year, etc.—is used. What is important is that sine and cosine transforms have mathematical tractability that the angle lacks.

Because so many people don't know about periodic regression (or sines and cosines), you are likely to encounter some reviewers who, despite their expertise in other areas, regard periodic regression as wizardry, or worse. Don't be surprised to find an occasional (and typically anonymous) one who will, rather than face the methods, emptily dismiss the work of which it forms a part. You are a sorcerer! As a reviewer, follow your responsibilities, and expect editors to follow their responsibilities to hold you to yours, as in the ethical guidelines† of the CBE (Council of Biology Editors, now Council of Science Editors) style manual (1983). As an author, be patient; but support journals whose editors hew to those responsibilities.

† a substantial CSE ethics excerpt is available (Sept. 2008) at
  http://www.nefsc.noaa.gov/publications/crd/crd0301/#app3

Enjoy the book. Statistics is, in the best sense, a common language that crosses disciplines and topics. It is a tool of objectivity that helps us abandon a blinding attachment to prior conclusions; a tool to dismiss a prime source of trouble and waste. We may be bold with our statistical methods if we are equally open about them.

SHORT LIST: MAIN SYMBOLS AND SPECIAL NOTATION

*   asterisk indicates multiplication (*standard in computer programming and Excel; equivalent to symbol* $\times$).

**,^   a double asterisk or a caret indicates exponentiation (*also standard*), as does placement of exponent in superscript.

`         the grave accent (`), as in sin`$x$, marks proper conversion of following term to an unspecified standard angular format (rads, deg, etc.) before taking transforms, such that sin`x can be read "proper sine of x" or "sine of properly converted x". (See k, below.) The conversion so indicated is multiplication by $k_{standard}/k_{as\_measured}$. Thus, generally, sin`DOY etc. equates to sin(DOY\*$k_{standard}/k_{as\_measured}$); if using radians for calculation, sin`DOY=sinR`DOY=sin(DOY*$2\pi/365$).

R`       R followed by ` (subset of `) marks proper conversion specifically to radians (rather than other standard format).

(sin,cos)   the pair of transforms, as in (sin`$x$,cos`x), is a circular variable expressed in trigonometric format, not a parenthetical expression.

'   the prime (') marks provisional estimates.

"   the double quote or double prime (") marks the coordinate system implied by any angle: (x'',y'')=(sin`$x$,cos`x). It helps keep ideas clear and helps keep track of transformed data.

*See **Notation** (Sec. 13) for more detail.*

DOY   day-of-year; 0≤DOY≤365. See DOY table, Sec. 16. The phrase "Julian Day" is incorrect for this meaning (Sec. 5.3.1).

k the number of units in one complete cycle, as measured (e.g., a day could be k=24 (hours), or k=1440 (minutes), etc. Standard angular format radians has k=2π, degrees has k=360.

m, y, d, h, min, s, when used as units mean meter, year, day, hour, minute, second.

*See **Glossary** (Sec. 14) for more symbols.*

1 INTRODUCTION

1.1 WHY BOTHER?—BIOLOGISTS NEED TO KNOW ABOUT CYCLES

Biology is dominated by cycles. Organisms respond to external cycles (like time-of-day, stage-of-tide, time of year, phase of moon) by cycling their own activities, movements, feeding rates, growth rates, etc. Abiotic variables like temperature or rainfall can also respond to the same cycles.

Periodic variables are those that cycle according to external cycles; external cycles typically are circular variables.

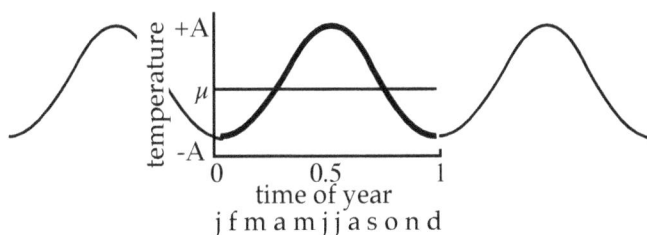

Figure 1-1. A familiar example of periodic variation: temperature is a *periodic* variable with respect to time of year, a *circular* variable. With circular variables the zero is arbitrary, which is why it's okay for New Year to be different in the Western, the Chinese, or other calendars. Therefore, the graph's X scale could be shifted right or left without loss of legitimacy.

Daily, tidal, lunar-monthly, and seasonal cycles (Figure 1-1) are well known in temperature, etc., and they exist in biological variables also. Only rarely, however, are they properly attacked in analysis. Too often, a graph of raw data is presented and a subjective (verbal) interpretation is given with no statistical (objective) support. Sometimes we'll see an ANOVA that—ignoring sequentiality—simply asks 'are some of these months different' and gives a yes or no that's inherently unreliable, because the nature of the relationship between months has been improperly presented to an inappropriate statistical procedure. It is in any case generally a bad thing to treat a continuum as a series of categories (linear data as nominal, see Table 3-I), because it amounts to over-rounding that loses potentially valuable information; if an analysis is available for continuous data, there is no justification for converting it to categories, nominal scales, bins, and so on.

As biologists we typically acknowledge cycles verbally but not quantitatively—in short, we pay them lip service. This little book seeks to correct that.

2 ANALYSING CYCLES

Cycles are just a bit tricky: times in a cycle are analogous to angles. You can't simply average angles as you would fish or gerbil weights. For example, take the arithmetic mean of the two northerly angles 359° and 1° and you get 180°, which you know is nonsense because it is due south. (Throw in 0°, due north, and try again!) Pigeons fly off in twenty directions, and the average direction is not the arithmetic mean. So, the arithmetic mean is meaningless for values in any cycle, whether days in a year, times in a tidal cycle, or times in a lunar month. For that reason, you cannot use these variables *directly* in analysis. Before averaging or using them in analysis, you need to transform, or decompose, them into their sine and cosine components.

> *Thus, the transformation of circular variables is the key to analysing them. We'll see: that each circular variable becomes two variables in the analysis, that we shouldn't separate them, that in transforming them we think of an (x,y) coordinate system—which we label (x'',y'') (Sec. 13.3) to avoid confusing this Y'' with the Y that is your dependent variable.*

Seasonal, tidal, or lunar effects may often be our primary interest; but even if we are more interested in some other variable, inclusion of the circular variables can improve detection and resolution of other effects. Therefore, because of the importance of cycles, it is worth knowing (i) that they can be handled and (ii) how to handle them. Circular variables include compass directions, times of day, times of week, times of year, times within the lunar month, and so on.

Not only is biology dominated by cycles, it is also generally complex rather than simple. Single-variable explanations (analyses) are often wasteful, and can be misleading, compared to analyses that take into account multiple effectors, predictors, independent variables. Generally, if 3 independent (X) variables (of any kind) have effects on Y (dependent variable), a better description of each effect is had by analysing them together—and a worse description by analysing each in isolation without reference to the others (sadly, this is often done). We rarely include more than 10 variables: practicality and data limitations often require us to structure our analyses more simply than that, i.e. ignoring some smaller effects—or those we expect are minor—and focusing on a smaller number of major effects. Sometimes we're limited in the data we can collect, but there's no excuse for the frequent neglect to record the date and time of each observation.

Underlying cycles often affect, and confound, the dependent variable of interest. A regrettably popular way to deal with that is to *attempt* to constrain observations to the same stage of a cycle, but it is problematic. Firstly, it is logistically costly and often impossible if more than one cycle is involved—e.g., to sample at the same time of day and time of tide reduces you to one sampling opportunity per month (yes, people have done this!). Secondly, the inferences can be meaningless outside such restricted times of the cycles. Thirdly, the opportunity to describe periodic effects is lost. Instead of attempting to leave cyclicity out of data, it is easy and far better to incorporate it into the analysis. That can more properly partition variation in observed quantities (*Y*) to any combination of linear and cyclic or directional variables (*X*). This is a generally unexploited opportunity to use circular *X* variables to *account for* or *remove* the effects of cyclic variation in *Y*, so that the remaining unexplained variation will be more readily explained by, or associated with, variation in some other variable.

Even where the primary interest is the relation of *Y* to some non-cyclic *X*, if the cycle also affects *Y* then including it (in a periodic regression) will likely improve resolution of the variables of interest. It's also possible to remove the effect of periodic variation, so that analysis of remaining variables will be improved. As a bonus, you get information about the effects of cycles on *Y*. For many questions the relation of *Y* to the cycle itself is the more important question. Thus, whether the objective of the investigation is the relation of *Y* to a cycle, or the relation of a *Y* (that is also affected by cycles) to some other type of variable, the proper methods of handling circular and periodic variables will be an asset.

1.2   GETTING STARTED

Circular is a data type; circular variables cycle back to the same point *by continuing to move in the same direction* and encounter each value once per cycle. Typical circular variables are well represented by a circle (Figure 1-3A). **Periodic** is not a data type, but a behaviour of linear (Table 3-I) *Y* data that vary according to a circular *X*. Periodic variables (like the sine or cosine transforms of the circular variable) reverse during their cycles, and typically encounter each value twice per cycle (except for the maximum and minimum). Typical periodic variables are well represented by a sinusoidal curve (Figure 1-3B).

Some subtlety is unavoidable with circular data. Angles are used in two senses: direction and rotation (Figure 1-2). Rotation reflects change in direction (an object can be rotated through a series of orientations). Rotation is sometimes used in a linear rather than circular sense (e.g. degrees of rotation by a wheel as a vehicle moves a given distance) and in that linear sense simple addition, multiplication, averaging pertain. Directions however are circular data and circular sense, and cannot be as simply averaged, as we will see.

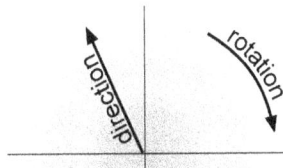

Figure 1-2. Direction vs. rotation; both can be described as angles, but treatment of the angles will depend on the purpose.

1.2.1. TWO DIMENSIONS: *X*

We typically think of the circular *X* (like an angle or hour of day) as a single dimension, but actually any circular variable, any cycle, occupies (at least) two dimensions (Figure 1-3). Proper handling of the circular, essentially two-dimensional, *X* is the key to analysing cycles.

Directional measurements like wind direction require careful handling, even for simple measures like the average direction among several measurements. Most calculations—from averaging angles to analysing for *periodic variation*, for example the change in temperature (*Y*) over a year, day, etc. (*X*)—require sine and cosine transforms of circular (or directional, cyclic) measures.

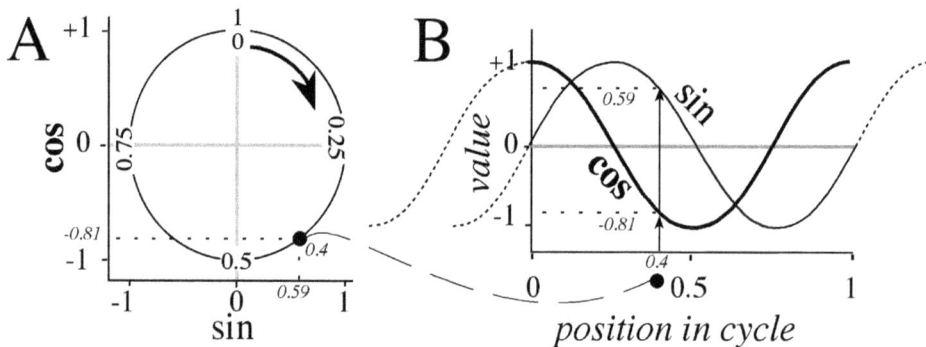

Figure 1-3. Cycles occupy two dimensions. **A**: Sine and Cosine (both periodic, Table 3-I) express angles 0 to 1 on a cycle (circular, Table 3-I). 0 to 1 on the cycle can mean 0*k* to 1*k*, *k* being the number of units in a full cycle, e.g. 0 to 360 degrees, 0 to 2π radians, 0 to 24 hours, etc. **B**: Sine and cosine vs. cyclic scale corresponding to position in the cycle. The nominal zero of a cycle is arbitrarily defined, with little real meaning; it's just a reference point. We can re-define or move the arbitrary zero according to what is most convenient for analysis. Note that this book favours the azimuthal system (Sec. 3.2.3); for the polar system we would place the sine on the *Y* axis and angle 0 on the right limb of the *X*-axis on panel A.

Directional variables are *circular* because as the compass pointer (or clock hand, etc.) proceeds in one direction through increasing angles, it will eventually return to the same position. (That's in cyclic or angular format, Table 3-I; but the sine and cosine transforms—the two dimensions—of the circular variable are [each] periodic, showing reversals and encounters of the same value twice per cycle, as in Figure 1-3.) The circular scale zero is arbitrary; for example, there is no real sense in which south>east.

Sine and cosine transforms, the key to analysing cycles, can be treated as a coordinate space, as in Figure 1-3A; that figure becomes the base of Figure 1-4 where it supports visualising *Y* over an annual cycle. The notation (*x″*,*y″*), meaning (sin`*x*,cos`*x*), identifies this coordinate space (see Notation, Sec. 13 for these and other special conventions).

1.2.2. THREE OR MORE DIMENSIONS: *Y* PLUS TWO PER *X*

The phrase "circular statistics" involves circular variables (*X*), and also phenomena, usually the *Y*, that respond in a *periodic* way to a (circular) *X* variable. Circular statistics includes procedures addressing *X* alone, and addressing both *Y* and *X*. Our main interest is where *Y* varies periodically with a circular *X* like a date in the year or a direction, but to analyse a periodic *Y*, we first must understand the circular *X*.

Even where a behaviour with respect to the cycle is not the main motivation for a series of measurements, the accommodation of cyclic variation in the analysis is far preferable to restricting sampling to a single stage of the cycle. That's because accommodating cycles incidental to the analysis can improve resolution on the effects we're primarily interested in—just as accommodating additional variables (*X*2, *X*3...) improves our analysis of the first independent variable (*X*1).

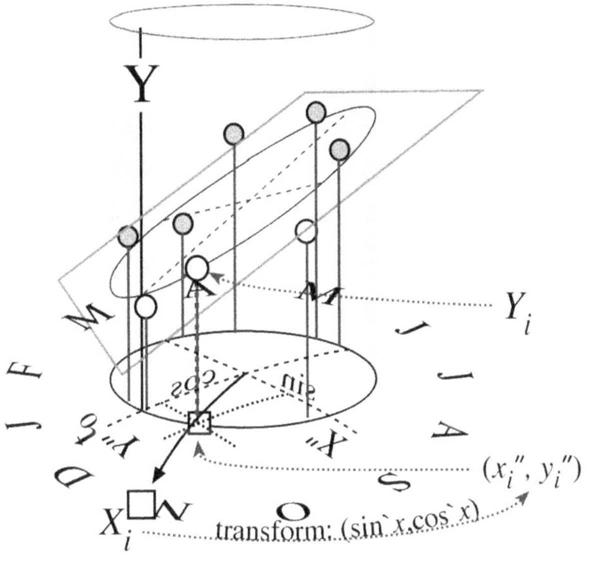

Figure 1-4. Periodic variation of *Y* over an index cycle *X* (e.g. a year). The base is a circular index variable (day, year, etc.), and height represents *Y*. An example point, angle X_i, represents a time in November; proper sine and cosine transforms identify the corresponding point (x_i'', y_i'') in the (X'', Y'') coordinate system. Thus the base can represent an index cycle as angles or (generally) as proxy axes in an $(X'', Y'') = (\sin\grave{\ }x, \cos\grave{\ }x)$ 2-D coordinate system (Sec. 13). The cartooned data have a trend approximated by a flat plane intersecting the cylinder; this is sinusoidal variation, the simplest possible periodic form. The null hypothesis would be a horizontal plane. Data are on the surface of the cylinder because the circle represents time in the cycle, but positioning inside or outside the circle could represent an additional variable or dimension; compare the polar plot (Sec. 6.5.2), which would represent *Y* as distance from center.

Variables like temperature are not essentially periodic, they're just scales. E.g.: the temperatures of steel, ice, soup or chocolate are not periodic. But temperatures of air in Rome, or of the Caribbean sea surface, can be periodic (can cycle) with respect to season, time of day, etc. Thus, it takes two to tango: whether a Y is periodic depends on what X is considered. Periodic systems (Figure 1-4) are thus (at least) 3-dimensional: one for the Y and two for the circular X. This is the "cylindrical" situation, and is analysed by periodic regression (Sec. 6).

We need ways to analyse this kind of situation, and that's what this book is about. Not only can periodic regression do that, but it is flexible enough to handle additional variables, whether circular or linear, to make usefully complex models.

1.2.3. PLANNING A STUDY: ANALYSIS AND SAMPLING

Everything affects everything. Analysis should include not merely those X variables we are interested in, but any others that plausibly affect Y. But we cannot measure everything, so we rely on the assumption that most of the effects are small, and we assume that we can identify the most important things affecting what we are interested in (that is often the weak point in a research plan). The variables we address—or ignore—are:

a.   our Y;

b.   the Xs ($X_{primary}$) of main interest;

c.   (optional) Xs ($X_{secondary}$) of secondary interest: we analyse them together with the main Xs;

d.   (optional) Xs ($X_{held\_constant}$) of tertiary interest: we cannot or do not wish to analyse them but we hold them constant during our measurements;

e.   (ignored) all the other Xs ($X_{ignored}$) in the universe, because we think or hope their effects are small.

Items [a] and [b], and sometimes [c], account for most analyses. Only rarely are cycles included under [b] or [c], and often they are improperly handled (seasons or months treated as categorical variables, etc.).

Where [d] is a cycle (a circular variable), it's often ignored because the researcher (or his supervisor) doesn't know how to analyse one. Students are often advised to restrict sampling to a narrow range within the cycle to avoid having to actually (horrors) analyse one. The hope (never tested in that situation) is that its effects are negligible; but if the cycle is influential enough to think about, then treating it under [d] means at best that the findings are meaningless outside the tested range.

Likewise, ignoring temporal variables under [e] can lead to mistaken conclusions. Cycles should be included under [c] unless there's clear reason to think they have negligible effect. This reminds me of Leland Wilkinson's quip about a constant, or intercept, in a regression model: *"if you are not sure you want a constant in your model, you want a constant"*

(Wilkinson 1987). In that vein: if you are not sure whether you want to include a cycle in your model, you want that cycle.

Periodicity should be accommodated in analyses, rather than excluded from data.

Restricting variation in circular variables—like time of day, stage in tide, or time in lunar month—has costs:

[1] the findings of the study are meaningless outside the limited cycle times sampled (e.g. spring high tide, or mid-day, etc.);

[2] the resolution on the variables of main interest is diminished (because the variation due to the cycle is unattributable);

[3] nothing is learned about the cycles; and

[4] logistical difficulties in the study are amplified; e.g. researchers must race from station to station to complete sampling within a short time of the daily cycle, thus limiting the workable number of stations and possibly reducing the quality of work.

The worst of those consequences is that the findings are meaningless outside that range. Some workers try to compensate by replication, but although replication does no harm it is an expense, and philosophically replication within a range does nothing to legitimise extrapolation beyond it.

How best to cope with cyclic variations? We should aim to fairly cover all cycles. If using regular sampling, ensure it will sample at least 6 to 8 different points in the shortest cycle without leaving a large un-sampled gap. Evenly spaced observations are generally considered desirable in statistical models; but on the other hand they can introduce an artificial frequency and, particularly in long studies, regular sampling is often logistically difficult or unachievable (Sec. 6.4.6). Instead, sample at random times, or possibly at haphazard times according to logistics and efficiency (this is like assuming that the haphazard factors at any particular time add up to something approximating a random number). It is easy to do ... even easier than constraining data. All it requires is: sample when possible, collect enough data to compensate for lack of symmetry. Always record the date and time of each observation: these can later yield (Sec. 5.3.1) the annual, daily, lunar and tidal cycles, and those can be appropriately transformed for an analysis that can say far more than an analysis of constrained data could.

Likewise for other variables: if you have ready access to them, record them. You get better resolution on your question, your study is more robust, and you discover more with the same effort.

2 ROADMAP: WHAT CIRCULAR STATS DO YOU NEED?

The circular stats routine you need depends on what kind of data you have and the kind of question you want to answer. You may be facing this question having already obtained your data; or you may be more fortunate, or wiser, thinking about this before you design your study.

Here are some approaches matched with easily-visualisable hypothetical examples.

A "periodic" Y is one that varies with respect to a "circular" X. The situations presented here involve a Y that is linear (or capable of being made or treated as linear), *and at least one circular X*. There can be additional Xs, and they can be circular, linear, or categorical.

2.1 SINGLE-VARIABLE PERIODIC STATS (YOU NEED: CIRCULAR AVERAGE, RESULTANT)

You have: one *variable* only (and in different circumstances you could call it X or Y), and it is circular (i.e. angles; directions; times of year, lunar month, day, tide etc.). You want: an average appropriate for a circle or cycle e.g.:

A: mean of a number of directions (e.g. 103 ants are released from a central point, and the direction is taken from their positions after 30 seconds); *question is* "what's the mean direction?"

B: mean of a number of dates (e.g. dates of hurricanes occurring in a defined area); *question is* "what's the average date of a hurricane" (remember, results in this situation will be meaningless unless you have sampled, or observed, evenly throughout the cycle)

Firstly, do not apply an arithmetic mean! Use the mean angle (Sec. 4.1). Graphically, visualise this situation as a number of points on the unit circle (see Sec. 3.2.3, Table 4-I, Figure 4-2).

Provided the samples are chosen in a way that does not bias them to one part of a cycle, useful calculations include the mean vector, the resultant vector, the resultant angle, and the mean angle or circular average. These can be plotted if desired, and significance of directionality or phasing can be tested using the Rayleigh test (Sec. 7.1).

If the sampling was biased to one part of a cycle, under the assumption of the cycle form (e.g. sinusoidal) the data may still be informative through periodic regression.

2.2 MULTI-VARIABLE PERIODIC STATS (YOU NEED: RELATION OF *Y*-VARIABLE TO CYCLE)

We can call a linear *Y* periodic if it shows a significant response to a circular *X*. The text is the main part of this section; the equations are supplied to help visualise the form of the analysis. See notation, Sec. 13 or page xv, for marks `, ", and '.

2.2.1. IF CYCLES ARE OF DIRECT INTEREST

Periodic regression—one cycle (Sec. 1, 6.4.1): You have: one *Y* (linear), one *X* (circular). You want: relation of linear (*Y*) variable to a circular (*X*) variable. Each circular variable is always transformed into $\sin`X$ and $\cos`X$, and regression must estimate a coefficient for each. The form is like:

$Y = B_0 + B_1 * \sin`X + B_2 * \cos`X$, e.g.:

A: 103 ants are released from a central point, and their positions after 30 seconds are taken as distance (*Y*) and direction (*X*); *question is* "how is distance related to direction?"

B: photosynthetic activity (*Y*) is recorded at a number of times (*X*) of day); *question is* "how is activity related to time of day?"

C: harvest quantity (*Y*) of a certain fish at times of year (*X*); *question is* "how is harvest related to time of year?"

D: temperature (*Y*) at various times of year (*X*)

Graphically, visualise this situation as a number of different-length vectors plotted from the graph origin (e.g. Figure 4-1 "data"), or as points vs. some representation (linear, circular, sectional) of a cycle (as in Figure 1-4, Figure 6-5, Figure 6-6, or Figure 6-10).

Periodic regression—two or more cycles (Sec. 6.6): You have: one *Y*, several *X*s (circular). You want: relation of linear (*Y*) variable to (here, two) circular variables (*X*1, *X*2). Numbered *X*s are different variables, and observations can be indicated by subscripts like $_i$. The form is:

$Y = B_0 + B_1 * \sin`X1 + B_2 * \cos`X1 + B_3 * \sin`X2 + B_4 * \cos`X2$, e.g.:

A: 103 ants are released from a central point, and their positions after 30 seconds are taken as distance (*Y*) and direction (*X*1), and this is repeated at various times of day (*X*2); *question is* "how is distance travelled related to both direction taken and time of day?"

B: photosynthetic activity (*Y*) is recorded at a number of times of day (*X*1) and at a number of times of year (*X*2)

C: harvest quantity (*Y*) of a certain fish at times of year (*X*1) and at times of lunar month (*X*2)

It is difficult to present multiple cycles graphically. Use one graph per cycle to visualise (as above) this situation. Tabular output becomes much more important as the complexity of the analysis increases.

Bivariate plots of parts of complex systems should be seen as supplementary to tabular regression outputs, because the limited dimensionality of plots can obscure cycles that are truly present.

Periodic-linear regression—one or more cycles and one or more linear X variables (Sec. 6.6): You have: one Y, several Xs (some circular, some linear). You want: relation of linear (Y_i) measures to (here, two) circular ($X1_i$, $X2_i$) measures, AND a linear variable ($X3$). The form is:

$Y = B_0 + B_1 * \sin`X1 + B_2 * \cos`X1 + B_3 * \sin`X2 + B_4 * \cos`X2 + B_5 * X3$,

e.g.:

A: 103 ants are released from a central point at various times of day ($X2$, circular), and their positions after 30 seconds are taken as direction ($X1$, circular), temperature ($X3$, linear) and their weights are also taken (Y); *question is* "how is distance travelled related to direction taken and time of day, and to temperature† and the ants' weights?"

(† With temperature as an X as well as time of day,   intercorrelation with time of day would be likely, and your best path would probably be to first de-trend temperature ; see under Sec. 6.4.1.)

B: photosynthetic activity (Y) is recorded at a number of times of day ($X1$, circular) and at a number of times of year ($X2$, circular), and at various temperatures ($X3$, linear)

C: harvest quantity (Y) of a certain fish at times of year ($X1$, circular), at times of lunar month ($X2$, circular), and with various market values ($X3$, linear)

D: barometric pressure (Y) at various times of year ($X1$, circular) and various altitudes ($X2$, linear))

As above, visualise cycles separately, e.g. one graph per cycle, plotting the Y vs. X (linear). For complex regressions, interpretation must rely substantially on the tabular output.

Periodic-linear regression with a category or groups (Sec. 6.6): You have: one Y, and circular, linear, and categorical Xs. You want: relation of linear (Y_i) to a circular ($X1$), a linear ($X2$), and a categorical or nominal scale variable ($X3$, which may be addressed by 'dummy'[a] variables for its categories). Forms may be like the following, with coefficients B_0 for the intercept, B_1 & B_2 for the periodic components of $X1$ ($\sin`X1$ and $\cos`X1$), B_3 for the linear term, and B_4 etc. to address $x3$'s categories:

$Y = B_0 + B_1 * \sin`X1 + B_2 * \cos`X1 + B_3 * X2$
    $+ B_4 * X3(\text{II}) [+ B_5 * X3(\text{III}) ...]$, e.g.:

A: 103 ants are released from one point, and their distances (Y), and direction ($X1$, circular) taken after 30 seconds; this is done over a range of temperatures ($X2$, linear), and for two or more species

---

[a] For c categories in a categorical variable, $c-1$ dummy variables will be needed.

(*X3*=I, II, III, ..., *species*; categorical). *Question:* "is distance travelled related to direction taken (etc.) and to species (etc.)?"

B: blood pressure (*Y*) is recorded at over times of day (*X1*, circular) in patients of various weights (*X2*, linear), and this is done for two or more treatments (*X3*=I, II, III, ..., *treatment*; categorical)

This is a complex structure. We could use separate regressions for each category. We could also use dummy variables for each category, or first use regress the linear vs. circular components, then use ANOVA to analyse residuals by category (these assume responses or slopes are equal for each category). A similar approach would be ANCOVA.

As above, visualise cycles separately, e.g. graph per cycle, and consult the tabular output.

Comparison of phase in previously analysed cycles (you need: Rayleigh test, Sec. 7.1)

Comparison of phasing (with respect to a cycle) among populations can be done using the regression output coefficients for the cyclic components. The *Rayleigh test* (Sec. 7.1) tests for central tendency among several populations' indicated peaks (point estimates) -- you will usually need 5 or more populations (or peaks or directions) to use this test and you get an overall indication regarding the entire group. If your observations were biased to some part of the year, avoid using the Rayleigh test to evaluate a central tendency in any phenomena that are not restricted to the times you observed; otherwise, the Rayleigh test will be biased by your own presence and absence (but you may be able to use periodic regression in that situation, with the dates as *X* and some other variable as *Y*). I.e. "significance of the mean date of storm occurrence during my holiday" is a trap.

2.2.2. IF CYCLES ARE IMPEDING ANALYSIS

If cycles are not of prime interest but are interfering with analysis of another variable of prime interest, then periodic regression provides a way of accommodating the cycle to improve resolution of the process of interest; or it can be used to de-trend data to allow other analyses.

Often a researcher is interested in, say, a growth rate vs. size of the animal. But if experiments were done simultaneously, then changes in temperature/seasonal effects will contribute to the observed variation in *Y*. That seasonal contribution can compromise analysis of the effect or dynamic of interest. One approach to improve the analysis—and it has saved at least one study—is to conduct a periodic regression such as growth vs. DOY (i.e. growth = $Y = B_0 + B_1*\sin{}DOY + B_2*\cos{}DOY$) (see Sec. 13 for notation), then use the residuals as *Y* corrected for seasonality, to improve analysis of *Y* vs. (e.g.) weight.

3 BASICS: TRIGONOMETRY, CONVENTIONS, FORMATS, CONVERSIONS

Why do we even need Trigonometry? We need it because the best analogies available to us for temporal cycles are models based on angles, circles, triangles. Analysis of cycles absolutely requires a little basic trigonometry!

Key to periodic regression is that some dependent variable Y is regressed against one or more independent *index*† *variables*, such as time of day or season, that are circular and that therefore must first be converted into standard angular units so they can be represented in the analysis as their sines and cosines.

> †   Index variable: an X or independent variable (such as time or location) to which the observations (Y) are indexed. This is subtly different from an X that might *affect* Y in the sense that rainfall might *affect* crop growth. Time as an index variable can be either in the linear sense (e.g. if we are exploring a monotonic trend)—or in the circular sense and requiring conversion to (e.g.) daily or seasonal index and its angular and trigonometric formats. (I say "sense" because one index variable can give rise to several temporal circular and one aperiodic variables.)
>
>   The meaning of the index variable is often indirect. For example, seasonal effects are well known to exist, but we usually do not know or account for the intermediates by which the position of the Earth in its orbit around the Sun might affect Y. Those intermediates, e.g. the orbital path, inclination of the Earth's axis, the intensity of radiation, rate of re-radiation, atmospheric composition, etc. are a 'black box', paradoxically unknown and yet reliable in our human experience via proxies like time of year. The Earth's orbital position is associated with variation in the distance from the Sun to the Earth, which of itself could generate seasons, and the fact that seasons are out of phase in northern and southern hemispheres is due to the tilt of the Earth on its axis. The point is: when making inferences about the "effect" of index variables, we accept that the "effect" is indirect, and via any of many unaccounted and unspecified things that do change along with those index variables.

Day of year and time of day are examples of circular variables, because they are units which repeat over and over. The relation of linear variables to cycles is visualised in Figure 5-1.

3.1.1. INDEX AND CIRCULAR VARIABLES

Suppose a particular apple, developing on the tree at a particular date and time t, has diameter y. The y is measured, but the t is merely recorded. The t is generally not subject to measurement error, unless you are using a sundial. The variable t is best thought of as an index variable that can be readily interpreted, according to need, as points on multiple cycles: time of day, time of year, time of lunar cycle, time of tide, etc.

That single index variable t can be properly transformed to map one or more (Figure 5-1) cycles (circular variables) for use in an analysis.

3.1.2. SINE AND COSINE OF AN ANGLE

Many people think they hate trigonometry. We too easily forget the incredibly simple basis of sines and cosines. Here is a mental image that will help you remain calm: imagine (Figure 3-1) a 24 hour clock (with an hour hand 1 unit long); then, to get the sine and cosine of any time, separately measure (using the hour hand's unit) the horizontal and vertical distances from the clock's center to the end of the hour hand.

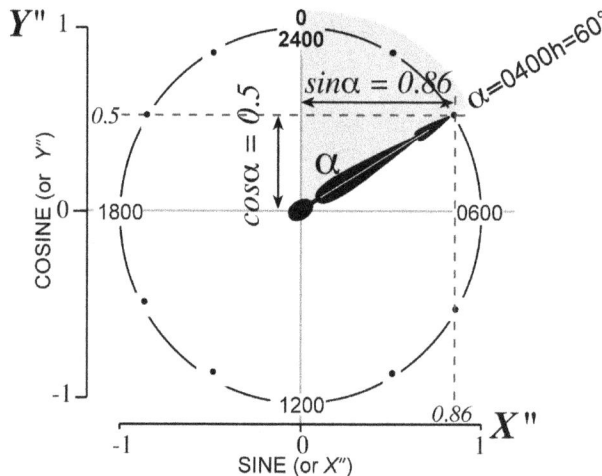

Figure 3-1. A cycle can be mapped in any units: here, a 24-hr clock, with radius = 1.0 (unit circle, see Figure 3-3). Time can be shown as an angle (α). Because radius=1.0, sine and cosine of α are measurable horizontally (x'') and vertically (y'') from the center (otherwise they would be x''/radius and y''/radius). The circumference can be 0–24 h, 0–365 d, 0–29.5 d (lunar), etc.

The take-home message—for trig-phobiacs to notice—is: *we got the sine and cosine without ever choosing whether to use degrees, radians, grads, etc.; so all of those choices are mere conventions that are all supposed to give the same result.* We could conceptualise this for any cycle.

Try to think of the (sin,cos) pair as being the *fundamental* and *universal* descriptors of an angle, and all those other conventions like degrees, radians, days-of-year, hours-into-the-tidal-cycle, etc., as, well, just conventions. It's easy to convert an angle from one kind of units into another, e.g. 0400h in a 24 h cycle equates to 60° (i.e., 360*4/24=60); we need such conversions to obtain the *true* (sin,cos) measures in the cycle of interest (daily, seasonal, etc.).

We can record circular variables, or the index variables from which we can get them, in various common formats like month-and-day, days-of-year, etc.. We can think of the circular variables as angles, but analysing them usually requires they be expressed in terms of sines or cosines.

Even the average has to be calculated a special way for circular variables (but it isn't difficult). While we can simply take the arithmetic mean of the weights of chocolate bars, or potatoes, or gerbils, or fish, and get a meaningful result, we cannot do that with angles. For example, try averaging wind directions of 350° and 10°: 360/2=180, a nonsensical result because approximately northerly winds cannot average to a net southerly wind. Clearly then, we need a different way to handle circular data (which are angles or can be treated as such). *The correct way to find the mean angle (Sec. 4.1.1) is—recognising that a cycle exists in 2 dimensions—to separately average the sines and cosines, then put them back together to get the mean angle.*

3.2   ANGULAR UNITS AND SYSTEMS

With circular variables you have a choice of *units* and *systems*. Most generally, a cycle contains k units; typical values for k are 360° (degrees), or 2π rads (radians, the usual angular unit in computing). Most calculators can be set for degrees or rads. Familiar natural cycles are 24h, 365d, 12.5h, 29.5d. You can make up your own arbitrary units, i.e. set k at whatever you want so each unit is $1/k$ cycles. This k is simple but tremendously useful, so please try to be familiar with it.

You also have choice of *visualisations* (azimuthal or polar angle; see Figure 3-2). Your results are the same whichever systems you use, you just need to stay in one system or know what you're doing. Should you choose the polar-angle system (which in physics and engineering is more conventional), take care in plotting, and in the quadrant correction (Sec. 4.3) used to obtain an angle from coordinates.

The *units affect how you calculate,* but the *system (Azimuthal or Polar-Angle) affects how you visualise.* I.e., a periodic regression (Sec. 6), or finding angles (Sec. 4.3), is done the same way whether you are thinking in the Azimuthal system or the Polar Angle system. That's because the tangent function is defined in terms of the angle's sine and cosine, or the opposite and adjacent sides of the right triangle, and thus is unaffected by which visualisation system is used. It will make a difference for making a polar plot, or any similar plotting of a vector.

3.2.1. AZIMUTHAL VS. POLAR ANGLE SYSTEMS (VISUALISATIONS)

This book favours the **Azimuthal** system unless otherwise noted, with the alternative being the **Polar Angle** system (Figure 3-2). These are perhaps best thought of as *visualisations*, because in calculation the angle, its sine and cosine, the arcsines and arccosines (sin⁻¹ and cos⁻¹) are all the same; only in the diagram is there a statement of where the zero is and the direction in which an angle increases. To quote Batschelet (1981), the choice of visualisation *"is irrelevant for the mathematical treatment and for the applications"*. I.e., only when we draw a diagram does it make any difference which system we choose.

In the azimuthal (like the compass, or a clock) visualisation, angles increase clockwise from the positive Y axis, corresponding to North on a compass. In the polar angle visualisation, angles increase counterclockwise from the positive X axis. The roles of sine and cosine are switched in the two visualisations: we plot an angle x as $(x'',y'')=(\sin`x,\cos`x)$ (Notation, Sec. 13) in the azimuthal system, but as $(\cos`x,\sin`x)$ in the polar angle system. Angles are typically represented by α [alpha] for the azimuthal system, and θ [theta] for the polar angle system; in the context of circular variables x is used, and of course x units in a k unit cycle can be understood as, practically, an angle.

As W.G. Warren quipped: *"so long as you know what you're doing, you can do anything you like"*.

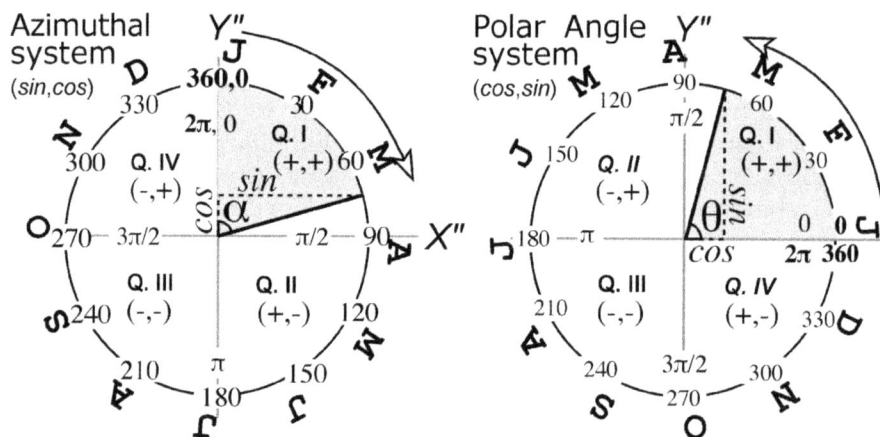

Figure 3-2. Azimuthal and Polar-Angle visualisations of the same angle. Any circular scale (rads, deg, DOY, months, etc.) can be represented. Both can be expressed on Cartesian (x,y) coordinates [here marked (x'',y'') (Sec. 13, Notation) to distinguish them from an independent variable X and a dependent Y]. **Azimuthal** visualisation (favoured in this book), oriented like a compass, measures angle α (alpha, in degrees, radians, months, etc.) clockwise from the positive Y axis. On the unit circle, coordinates $(x'',y'')=(\sin`x,\cos`x)$. **Polar angle** visualisation measures angle θ (theta) *counter*-clockwise from the positive X axis. Coordinates $(x'',y'')=(\cos`x,\sin`x)$. I.e., Azimuthal visualisation associates sine with x'' axis, but Polar Angle visualisation associates sine with y''. Quadrants (Q. I-IV) relate to angle; both systems number quadrants with increasing angles.

3.2.2. UNITS

To analyse cycles, you'll need to convert from linear units of X as observed to standard angular units—e.g. convert day of year to radians or degrees—then convert to (sin,cos). Only then can analysis be done.

Radians (rads) are the most popular angular units in computer programs, so conversion is usually from various units to radians. It's often easier to think in degrees, days, or whatever the original units were—it only requires a little notation (Sec. 13) to keep track of the units of transformed variables in spreadsheets. Pocket calculators can be set for degrees or radians.

Settings for degrees, radians, etc. need only match the k to which you converted your data. Azimuthal or Polar-Angle are settings in your mind and how you *visualise* your calculations—but this choice doesn't affect the relation of angle to sine, etc. (Your mental image being either azimuthal or polar angle would not affect your analysis; as said, azimuthal is the default here simply because it is an intuitive match with clocks and compasses, and time and direction are the most likely variables of interest for biology and medicine.)

3.2.3. CONVENTIONS: PREVIOUS AUTHORS

Bliss (1958; 1970), and Batschelet (1981) favour the polar-angle system of visualisation.

3.2.4. CONVENTIONS: THIS BOOK

This book favours the azimuthal (Sec. 3.2.1) or compass-like visualisation, because of its similarity to clocks and compasses, with which everybody is already familiar; the symbol α (alpha) is used for an angle. For compass directions $\sin(\alpha)$ is the east-west or X component, and $\cos(\alpha)$ is the north-south or Y component.

The notation R˙ (or ˙) indicates the proper transformation to radians (or other conventional units) allowing sine or cosine to be taken of a variable—e.g. sin˙TOD means the "proper sine of time of day", meaning TOD was, or is to be, converted to standard angular units before taking the sine. See Sec. 3.5 for calculation and Sec. 13.1 for notation.

Special symbols are given in the glossary, and special notation is described briefly (p. xv) and in detail (Sec. 13).

3.3 SIN, COS, TAN, AND AMBIGUITY OF INVERSES

Sine and cosine are not merely descriptors or associates of an angle; in fact the (sin,cos) pair is the most fundamental description of an angle.

Just as the angle doesn't change because of the angular units we use to describe it, whether degrees (cycle of $k=360$ units), or radians (cycle of $k=2\pi$ units), or whether in units like hour of day (cycle of $k=24$ units) or minute of day (cycle of $k=1440$ units) or minute of year (cycle of

k=525,600 units), etc., *the sine and cosine also remain stable descriptors of that angle.*

By definition, $\sin\alpha$=O/H, $\cos\alpha$=A/H. But the unit circle simplifies by keeping H = 1.0, so:

$\sin\alpha$ = **O**/1.0 = Opposite
$\cos\alpha$ = **A**/1.0 = Adjacent
$\tan\alpha$ = **O/A**
Handy: $\tan\alpha$ = $\sin\alpha/\cos\alpha$

Ambiguity: although any angle has only one sine, any sine has two possible angles. Likewise cosine.
  E.g. two angles have sine= +0.866; so although arcsin of 0.866 returns 60°, the angle 180°-60°=120° is also implied.

  That is why the arcsin and arccos functions return only one angle, but imply one more. To tell which is true, use the arctan method (see text).

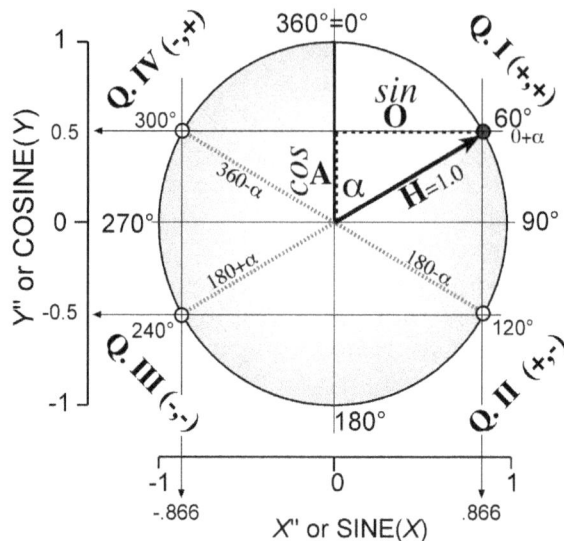

Figure 3-3. Trigonometry and ambiguities. An angle x or α can be visualised in a right triangle with components Opposite, Adjacent, Hypotenuse. The Unit Circle has radius = 1, a useful simplification because then sine = Opposite/1 and cosine = Adjacent/1. Sines (shown) are symmetrical across the X axis; cosines are similarly symmetrical about the Y axis (in the azimuthal visualisation). Tangents (ratio O/A) are symmetrical about the center, such that points directly opposite have equal tangent. (The X" and Y" notation (Sec. 13) denotes the dimensions sin`$x$ and cos`x to avoid confusion with usual Y, the dependent variable.)

The angular units are just an arbitrary series of marks drawn around the Unit Circle; useful, yes; but fundamental? ... no! The cycle is a fixed concept which we divide, depending on the application, into k parts—and k can be arbitrarily chosen convenient to the task: 360 degrees, 2π radians, 24 hours, or 365 days, etc.

Without knowing the k for the cycle in which α is expressed, we really don't know the angle. This is an aspect of the two-dimensional nature of cycles. We could think of $\sin(x)$ as $\sin(k,x)$, or sine of x given the cycle's period k—it might make things easier. Thus $(360,x)$ would be a full description of the angle, containing all the information needed for taking the sine and cosine: if our "sine lookup" table or function works with radians, it lets us say sin`$x$=$\sin(360,x)$=$\sin(x*2\pi/360)$; if the lookup table works with degrees then sin`x means $\sin(360,x)$=$\sin(x*360/360)$. The notation sin`x (Sec. 13) eliminates the need to write out all the units.

The Unit Circle (Figure 3-1, Figure 3-3) is a brilliant concept. It comprises the points that satisfy $\sin^2+\cos^2=1$, $\cos=(1-\sin^2)^{0.5}$, etc. The hypotenuse of an inscribed right triangle is a radius, therefore also 1.0 units long. The Adjacent side represents the cosine, and the Opposite side represents the sine. In other words, in (X'',Y'') coordinate space an angle α is plotted as $(x'',y'')=(\sin\alpha,\cos\alpha)$[†], and $\sqrt{(x''^2+y''^2)}$ is the length of any vector from $(0,0)$ to (x'',y''). [† in the polar angle visualisation, these are switched: $(x'',y'')=(\cos\alpha,\sin\alpha)$; Figure 3-2]

The quadrant (Q. I, II, III or IV) is apparent from the signs of the coordinates $(x,y)=(\sin\alpha,\cos\alpha)$ of an angle, and affords a quick check for plots. Quadrant is also important in calculating angles (Sec. 4.3.2).

A right triangle inscribed in the Unit Circle has sides Opposite, Adjacent, and Hypotenuse (Figure 3-3). Then, for the acute[b] angle α:

$\sin\alpha$ = opposite/hypotenuse = O/H

$\cos\alpha$ = adjacent/hypotenuse = A/H

$\tan\alpha$ = opposite/adjacent = O/A; this implies $\tan\alpha=\sin\alpha/\cos\alpha$
  (because we can multiply O/A by 1= 1/1= (1/H)/(1/H),
  giving (O/H)/(A/H), by which we see $\tan\alpha=\sin\alpha/\cos\alpha$). This
  is useful for recovering an angle from vector coordinates or
  its sine and cosine (Sec. 4.3, Figure 4-4)

$\{\sin(\alpha)^2 + \cos(\alpha)^2\}^{0.5} = 1$ (the exponent 0.5, for square root, has no
  effect in the special case of the unit circle, i.e. $1^2=1^{0.5}=1$; but
  as an important part of Pythagoras' theorem it is retained for
  its generality)

Numerous other equalities exist, and some illustrate the ambiguity of inverses, for example:

$\sin\alpha = \sin(180-\alpha)$ (degrees), or = $\sin(\pi-\alpha)$ (radians)

$\cos\alpha = \cos(360-\alpha)$ etc.

"**SOH-CAH-TOA**" is a useful mnemonic for remembering what ratios correspond to sin, cos and tan; the mnemonic stands for $\underline{S}$ine = $\underline{O}$pposite/$\underline{H}$ypotenuse, $\underline{C}$osine = $\underline{A}$djacent/$\underline{H}$ypotenuse, $\underline{T}$angent = $\underline{O}$pposite/$\underline{A}$djacent); these are diagrammed in Figure 3-3.

3.3.1. INVERSES AND AMBIGUITIES

Inverses of trigonometric functions are signified by the prefix *arc-*, as in "arcsine", "arccosine", "arctangent" etc. or "$\sin^{-1}$", "$\cos^{-1}$", or "$\tan^{-1}$". Read notation like "$\sin^{-1}(\alpha)$" as "inverse sine of α" or "arcsine of α", but ⚠ not $1/\sin(\alpha)$; it's notation, not a mathematical expression—the inverses of functions are not nearly that simple! It works as notation because, while $\tan(x)^{-1}$ means $1/\tan(x)$, $\tan^{-1}$ on its own is mathematically meaningless, just as $+^{-1}$ or $\sum^{-1}$ is meaningless; operators are not values, so operator^{-1} makes no sense except as a notation for an inverse function.

Although sine and cosine functions are not ambiguous, their inverses are. E.g., $\sin(\alpha)$ has just one possible result x, but $\sin^{-1}(x)$ has—like a square root—two possible results. Thus it is not simple to work

---

[b] Acute angle: $\alpha<90°$. Right angle: $\alpha=90°$. Obtuse angle: $90°<\alpha<180°$.

backwards from sine to angle; cosine is needed also. Any (sin,cos) pair implies one angle and it can be methodically recovered (Sec. 4.3).

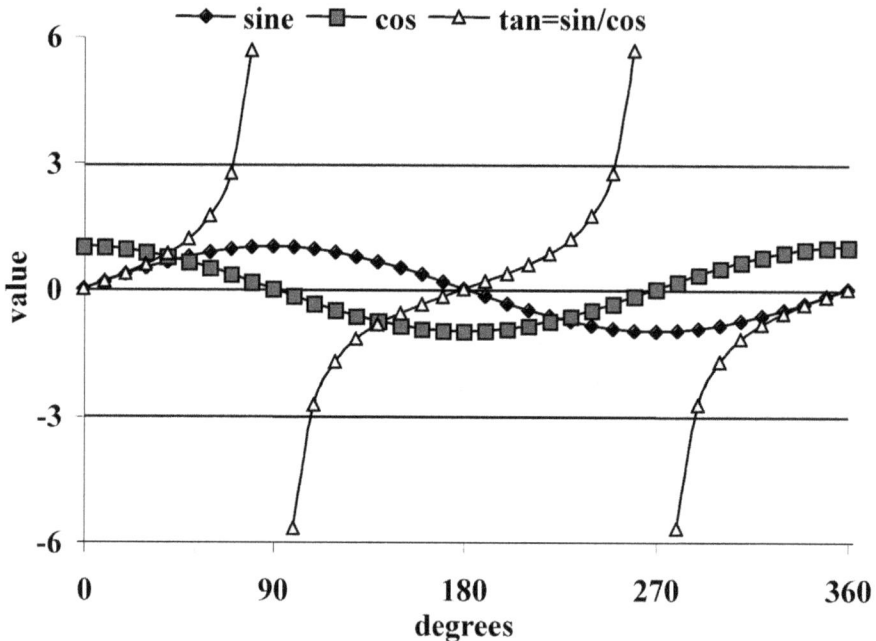

Figure 3-4. Periods of 3 trigonometric functions: sine, cosine, and tangent. Sine and cosine have period 360° or 1 cycle. Tan is discontinuous, with period of 180° or 0.5 cycle. Since Sine = Opposite/Hypotenuse, and Cosine = Adjacent/Hypotenuse, and Tangent = Opposite/Adjacent, Tan also = Sin/Cos. Tangent can be large or 1/large; this graph is vertically truncated.

WARNING: the ARCTAN, ARCSIN, ARCCOS functions ([INV][TAN], etc. on calculators) *assume* that $0° \leq \alpha \leq 90°$ (degrees) or 0cycle$\leq \alpha \leq$0.25cycle (any scale), and this means that extra steps are needed to accommodate angles >0.25 cycle. See quadrant corrections, Sec. 4.3.

Although sine and cosine are continuous, the tangent function is not. Their pattern over a cycle is quite beautiful (Figure 3-4). The fact that $\tan\alpha = \sin\alpha/\cos\alpha$ is useful for recovering true angles (Sec. 4.3) from just the sine and cosine, or vector coordinates, or regression coefficients.

We've seen a cycle X expressed on (x,y) unit circle coordinates that derive from the circular X variable; the coordinate y should not be confused with the dependent Y that is analysed with respect to the independent X. To avoid confusion, we will often write this $x=\sin`x$ and $y=\cos`x$ as x'' and y'' (see Notation, Sec. 13), the two dimensions that properly express a circular independent X as variables fit for use in analysis or calculation. (Take a breath.)

3.4   DATA TYPES, AND FORMATS OF CIRCULAR DATA

We occasionally have need to describe or name the types and formats of data we deal with. The data types here are substantially consistent with acknowledged (Zar 1984) data types. Direction and time are our main interest.

Time, *in the continuous sense*, is a variable like seconds, years, etc. since some zero (beginning of the universe, civilisation, etc.). Although we expect a true zero (so it should be ratio scale) we don't know when it was, and therefore use arbitrary zeros. Despite that, time is the major index for civilisation and conceptualising natural processes! Discrete or categorical time units are an illusion, at least at our time scales.

Time is in essence monotonic. Representations provide—directly or indirectly—circular time variables or cycles, but those do not inhere in time itself. Instead, natural cycles result from the regular motions of large objects through time and space, their orbits affecting their gravitational and radiative effect on Earth, thus generating periodicity in natural phenomena and giving rise to our annual calendar (a model of the Earth's orbit). Time thus has both linear and circular senses, so we can call it linear or circular depending on which sense is relevant.

I distinguish the circular data type (Table 3-I, Figure 3-5, Table 3-II) from interval data (Zar 1984) because [i] circular variables are essentially 2-dimensional (Sec. 1.2), unlike temperature etc. As well, circular variables contain subtleties not found in other data types. [ii] The zero in circular data is not arbitrary in the same sense as in scales like Celsius and Fahrenheit: 5°C>2°C, but there is no sense in which south>east, yet a rotation of 180°>90°, and [iii] the zeros of sine and cosine are real enough. Circular variables can be expressed in several formats, each having uses and limitations.

Converting amongst circular formats is key to analysing cycles. Trigonometric format (sin,cos) is fundamental, but we typically find a circular variable recorded in common format, which must be converted to trigonometric format for averaging, or for analysis. The conversion sequence is: ***common*** → [***linear*** →] ***cyclic*** → ***angular*** → ***trigonometric*** (Figure 3-5). E.g. (Table 6-II, Table 9-I) calendar date → DOY → [nDOY→] radians → (sin`DOY,cos`DOY).

Then, after analysis based on trigonometric format, we often need to convert results back to *cyclic* format for plotting, or to (further back!) a *common* format to discuss them.

Trigonometric format (sin`*x*,cos`*x*)

Trigonometric format: the sine and cosine of a circular measure or angle, or sometimes the coordinates of a vector (L*sin`$x$,$L$*cos`x) (Sec. 4.1, 13.2, 13.3). Trigonometric is the fundamental format for any circular measure; it's needed for averaging angles or dates or times in any cycle. For basic trigonometry see Sec. 3; for averaging and other calculations, see Sec. 4.

Table 3-I. Data types and their attributes. The note '*yn*' means the attribute can vary in the type. *Periodic* is not a scale but a behaviour (Figure 3-5); nevertheless, most periodic variables expected here would be linear scale. 'Index' refers to index variable (Sec. 3.1.1). See also Table 3-II.

DATA TYPE	ARBITRARY 0 ?	UNITS EQUAL?	CIRCULAR?	CATEGORY ?	OK AS INDEX?
CONVENTIONAL (ZAR 1984):					
RATIO SCALE: e.g. distance, count. Logical zero makes ratios meaningful.	*n*	*y*	*n*	*n*	*y*
INTERVAL SCALE: e.g. Celsius, Fahrenheit, date and time (in sense continuous, non-circular—this is a change from Zar's usage). Arbitrary zero makes ratios meaningless.	*y*	*y*	*n*	*n*	*y*
ORDINAL SCALE: e.g. size A,B,C, or 1,2,3, etc., ordered, ranked, such that B>A but B-A and B/A are meaningless.	*na*	*n*	*n*	*y*	*n*
NOMINAL SCALE: e.g. genotype, non-quantitative attribute.	*na*	*n*	*n*	*y*	*n*
OTHER TYPES & FORMATS RECOGNISED HERE:					
LINEAR SCALE: subsumes RATIO and INTERVAL data, and linear format (below) for monotonic sense of data like date, time or rotation, which otherwise have circular sense.	*yn*	*y*	*yn*	*n*	*y*
CIRCULAR SCALE: e.g. temporal cycles, angles, directions. Capable of expression on repeating scale; arbitrary zero. Units are equal in all formats except common.	*y*	*yn*	*y*	*n*	*~y*

FORMATS FOR CIRCULAR	EXAMPLE	UTILITY OF FORMAT
common	1240h, July 4, 1989, or 19890707.1240 (yyyymmdd.hhmm)	cannot directly plot or analyse
linear (time or rotation since arbitrary start; *see Linear Scale*)	549.528 (nDOY: days since 0000h, Jan. 01, 1988, neglecting leap)	can plot; can analyse only in monotonic sense, not used in analysing cycles per se
cyclic or non-standard-angular	184.527 (DOY)	cannot directly get sin and cos; can plot (e.g. Figure 6-5, Figure 6-7) but not analyse
[standard] angular	181.99° (degrees)	can directly get sin and cos; can plot (as with cyclic) but not analyse
trigonometric: (sin,cos)	(-0.0348, −0.999)	allows periodic analysis, average, etc.; can plot (e.g. Figure 6-6, Figure 6-10)

To illustrate conversion of formats *common* → *cyclic* → *angular*, and finally → *trigonometric*, take the common-format date "Oct 18". In cyclic format (DOY) it is 290; in angular format it is 4.992 radians or 286 degrees, and in trigonometric format it is (sin,cos)=(0.961,0.276). Table 6-II illustrates the conversions. As explained (Sec. 3.3), sine and cosine are unchanged, whether you use radians, or degrees, or you draw it out to scale and take the ratios (Figure 3-1; Sec. 3.2.3) of the opposite side to the hypotenuse, etc. DOY represents date as an angle (Oct 18 is day 290) in a cycle with $k=365$ units; most commonly it will need conversion to radians ($k=2\pi$), the usual angular measure in computing.

Angular format *x*: standard angle

Angular formats, e.g. degrees and radians, express circular variables in standard units from which sine and cosine (the trigonometric format) etc. can be directly taken. Angular format can be considered a special kind of linear or cyclic format. Calculation for most purposes will require decomposition into trigonometric format.

Addition is possible in cyclic and angular formats, just as if each day, degree, etc. was a fixed length of road that would be traversed by a wheel rotating one degree; all those degrees of rotation would sum to a total rotation which could equate to a distance traversed for a given diameter of wheel. They nevertheless have a problem: angles can't be simply averaged (average of directions 330° and 20° is not 350°/2). For analysis we usually need trigonometric format (Figure 4-2, Table 4-I).

There is a sense in which angles are linear (we can add them), and a sense in which they are not (we can't usefully apply the arithmetic mean). We therefore allow both cyclic and angular formats to be described as linear, if clearly in the context of circular variables.

Cyclic format *x*: non-standard angle

Cyclic formats are angular in concept (period *k* is acknowledged) but have non-standard units, meaning sine and cosine cannot be taken directly (unless we create a new table); e.g. day of year (*k*=365), hour of day (*k*=24) or even minute of day (*k*=1440), etc. Cyclic has a linear sense—and so for some purposes can be treated as linear format—but even in that sense it is unlike non-circular linear variables such as temperature. Taking sine and cosine (etc.) requires conversion (Sec. 3.5) to a standard angular format (usually radians); macros can be written to automate (Sec. 10.2) the conversion.

For dates, our preferred cyclic format is DOY (0-365) format (Sec. 5.2). DOY can generate nDOY (a linear format), useful in studies that span multiple years: nDOY is a count of days since DOY 0 of the first year in the study; usefully, sin`DOY = sin`nDOY (see Sec. 13.2 for ` notation). It's often useful to think of circular variables as angles having units 0-1 cycles, or *k*=1, e.g. DOY 127 as 127/365=0.35 cycles.

Linear format *x* (date etc.)

Date and time information straddles data types. It has two senses: circular (cyclic, angular, and trigonometric formats) and monotonic (linear scale or linear format, which need not acknowledge any *k*), as in Figure 5-1 and Table 3-I. Continuous time expressed on a interval scale is also a linear format, e.g. time elapsed since a reference point, for instance the (astronomical) Julian date with its zero in 4713 B.C. (Sec. 5.3.1).

Linear format (e.g. nDOY) is useful for plotting (e.g. Figure 6-5, Figure 6-7) and analysing the linear contribution of *X* to *Y*. For example, an analysis might accommodate periodic effects (*Y* vs. ... + sin`*X* + cos`*X* +

...), and linear effects (*Y* vs. ... + linear *X*). Linear formats can use cyclic or angular units, but *x* is in the range –infinity to +infinity; when 0≤*x*≤*k* linear is indistinguishable from cyclic or angular formats, which thus have a linear aspect. Linear is as much a sense as a structure.

•RATIO   (zero logical, ratios true)
e.g. length, weight, number (count), °Kelvin

0   1   2   3   4   ...   X

•INTERVAL ("0" arbitrary, ratios unreliable)
e.g. °Celsius, °Fahrenheit

"0"   1   2   3   4   ...   X
-2   -1   "0"   1   2   ...   X

•CIRCULAR (zero arbitrary, scale 0 to *k*, ratios unreliable)
e.g. month, DOY, degrees, radians

COMMON *format (month (of y,m,d)) as for date:*
... M   J   J   A   S   O   N   D   J   F ...

4 mathematically proper formats:

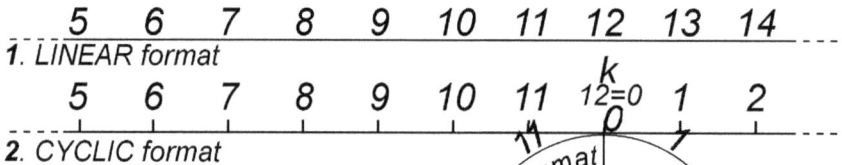

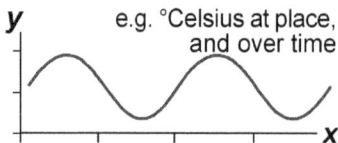

5   6   7   8   9   10   11   12   13   14

1. LINEAR format

5   6   7   8   9   10   11   12=0   1   2

2. CYCLIC format

y   e.g. °Celsius at place,
and over time

PERIODIC is not a type but a *behaviour* of a variable with respect to *x*. Sin(*x*) and cos(*x*) are also periodic variables with respect to *x*.

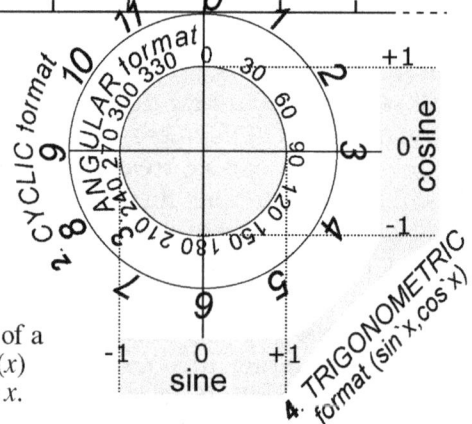

Figure 3-5. Key data types: *ratio, interval* and *circular*. Circular formats: common, converted to linear or to cyclic (non-standard-angular, e.g. DOY), then angular, then trigonometric (sine and cosine); compare Figure 5-1. *Periodic* is a behaviour, not a type; sine and cosine are also periodic variables. Except in the monotonic sense of time or rotation, ratios should be avoided in interval or circular data. See Table 3-I.

The linear sense of angles is useful in the context of total rotation, e.g. the number of rotations of a wheel while travelling a distance; typically, such situations show *x*>*k* or *x*<–*k*.

More subtleties: common or linear formats of date lose their multi-cycle information when converted to cyclic, angular, and trigonometric format. For example, it is simple to convert nDOY to DOY (keep removing 365 until 0≤*x*≤365), but information is discarded in the process, so reversing the conversion requires a source of the multi-cycle information that DOY cannot carry. Converting common format date to nDOY requires first converting to DOY, then obtaining the year from common date, subtracting the reference year you have selected as the zero of your nDOY, mutiplying the result by 365, and adding that to DOY.

Common format *x*

Common formats are often not mathematically proper, lacking a zero (even an arbitrary one), like the day of month. Common formatted time used in the circular sense requires conversion (Sec. 3.5) to cyclic or standard angular formats for plotting, and trigonometric formats for analysis.

Circular variables in common format can be converted to linear, cyclic, angular, and trigonometric formats. All but common formats can be mathematically proper (zero exists), but they aren't equally suitable for calculation.

Table 3-II. Circular formats. To identify format of an example, follow this key from the top down—from the most fundamental to the most arbitrary.

IF *X* appears as:	Then circ. format is:
paired units $(x'',y'')=(\sin`x,\cos`y)$ or non-unit vectors $(L\sin`x,L\cos`y)$, etc.	**trigonometric**
$0\le x\le k$ in standard angular units (rads, deg., etc.)	**angular**
$0\le x\le k$ in non-standard units (hours of day, etc.), in circular sense	**cyclic**
$-\infty\le x\le+\infty$, and *x* often <0 or $>k$, in standard or non-standard units, OR as angular or cyclic ($0\le x\le k$) but used in a linear *sense* like rotation or time since some arbitrary start	**linear** format or data type (a sense as much as a format)
x in mathematically improper format like °,m,s; d/m/y; yyyymmdd.hhmm; etc.	**common** (see Sec. 5.1.2)

3.5   CONVERTING CIRCULAR VARIABLE FORMATS

Converting calendar dates from common format to cyclic, linear and angular formats (usually radians, to obtain sine and cosine) is covered in Sec. 5.2. We can also convert back to common formats for communication. This section focuses on conversion amongst circular variables already in cyclic or angular form.

3.5.1. CONVERTING AMONGST CYCLIC OR ANGULAR UNITS

Natural† cycles (Sec. 5.2)require conversion to standard units like degrees ($k=360$), or radians ($k=2\pi$), to enable taking sine and cosine.

†Natural cycles—e.g. day ($k=24$ h, $k=1440$ min, etc.), or year ($k=365$ d)—are those with plausible biological meaning. I.e., we could say an hour is mathematically a cycle with $k=60$ minutes, but an hour is not a natural

cycle, it is an arbitrary fraction of one; the day could as easily have been divided into 10 'hours', etc., with equally little biological meaning.

The period of a cycle is expressible as k angle-like units. Conversion is simply by scaling. If given a common format date, see Sec. 5.3.1 or Table 16-I and get it into an equal-unit format before scaling. Generally, to convert an X value from a system of k_1 units to one of k_2 units,

$x_{k2} = x_{k1} * k_2/k_1$

Try taking the sine of 0400h, 4 a.m. in the day. Sin(0400) would be nonsense because our computer program expects x in standard units. Notation (Sec. 13.1) is thus needed to track conversions. The grave symbol (` ) indicates conversion to unspecified standard units; it lets us declare a conversion without specifying units. Radians are standard angle units, so R` further specifies that we intend to work in radians. Using the notation for our example:

sin`[0400h] =sinR`[0400h] =sin[2*π*(4/24)] =sin[2*π*0.1666...]
=sin[1.0471976...] =0.8660254; likewise, cos`[0400h]=0.5]

Whether via radians or degrees, the (sin,cos) result is fundamental (Figure 3-1). Both ` and R` represent ratios:

` = $k_{standard}$ / $k_{as\_measured}$ or, more specifically for radians:

R` = $2\pi/k_{as\_measured}$

Thus, to convert hour of day (0-24h) to radians, R`=$2\pi/k_{as\_measured}$ =$2\pi/24$h =0.261799 rads h$^{-1}$. To convert DOY to radians, R`=$2\pi/365$d =0.017214 rads d^{-1}. Any such conversion is easy to work out.

To quickly check conversions, try 0, 1/4, 1/2, 3/4 cycle in the original units and take sine and cosine; you should get 1s and 0s[†]. By watching what quadrant an angle is in, you can easily check that the signs of the sine and cosine are correct for that quadrant. An easy error is the angular system setting on your calculator.

[†]In spreadsheets however, although sin(180°)=0, and depending on cell format[††], a formula like "=SIN(180*2*PI()/360)" may return a value like 1.22461E-16, i.e. not quite zero. These tiny errors reflect the limitations of binary math, and such values should be reported as zeros.

[††] if no format is selected, the number usually displays in "E" notation, "E" meaning "times 10 to the power of"; if the 'number' format is chosen and some number of decimal places (less than, here, 16) is chosen, then these values will display as 0.000....

For notation and calculation of harmonics, see Sec. 13.4 .

4 KEY TOOLS: FOUNDATIONS OF CIRCULAR STATS

Directions (of growth, flight, etc.) and timing (on daily, seasonal scales, etc.)—all important in biology and medicine—involve circular variables. We may need to average or combine n directions or angles, to find the resultant or mean vector of n vectors of equal or varying lengths (L), or to find the mean angle (the circular average) of n angles.

Circular variables, however, do not sum or average like measures on linear scales. The average weight of oranges is $1/n$ the total weight of n oranges; not so with circular variables. Circular variables behave unlike linear variables, and therefore require special handling in calculation.

For example, consider two wind measurements, one 5° to the left of North and the other 5° to the right. Their average should be zero (i.e. North), but these measurements would have been recorded as 355° and 005°, and the simple average of these comes out to 180°, or South. *Clearly, that 180° is wrong.* This is (Fisher 1993) "the notorious cross-over problem familiar to meteorologists". *There is something peculiar about circular variables and we need a different way to deal with them.*

This section deals with components for calculation, the summary vectors (resultant and mean), the mean angle (often called the circular average); it deals with recovering angles from components; with rotating angles and finding the smallest difference between two angles.

An appendix (Sec. 10) shows how to program some measures in spreadsheets.

4.1 COMPONENTS, VECTORS, AND SUMMARIES

As explained by Zar (1984) and others, with circular variables *the zero is arbitrary: there is no true zero.* A compass heading of North is no less than a heading of East or South, even though North is identified with the heading 0° = 360°, while E, S and W are 90°, 180° and 270°. A cycle's end is like its beginning: $\sin(0°) = \sin(360°)$; and $\sin(x) = \sin([1\text{cycle}]+x)$, etc. ... no true zero.

Because a cycle occupies two dimensions, most calculations require decomposing a circular variable into its two components.

See (Table 4-I) how unreliable the arithmetic average is for circular variables: suppose we had data on the directions taken by a fish, bird or insect when released, and we want to know what direction the group tends toward. First, we calculate the arithmetic mean of the data, and, of course, we get a number (194.29°); but is that correct? Let's plot it (Figure 4-2): we find that the result is opposite to the actual data; clearly the arithmetic mean is unreliable for circular variables. (The arithmetic mean's errors become small only as the data become close together and provided the range does not cross the zero.)

The (proper) average obtained by the component method is analogous to graphically chaining together (Figure 4-1) n vectors with directions α and magnitudes L†, e.g., daily motion of some object drifting with the wind. This gives the resultant vector (RV, e.g. net total motion) and dividing its length by n gives the mean vector (MV, e.g. net average motion). (†L is ignored or set to unity if mean angle MA (below) is desired.)

Table 4-I. Mean angle (MA) of n angles or unit vectors. The 'Direction' column contains angular data (B4:B10). Here, k=360 (360°). 'WRONG' shows simple average, inappropriate to circular data. Resultant vector (RV) coordinates are the sums of sine and cosine components of constituent angles; divide these by n for mean vector (MV) coordinates. The arctan method (Sec. 4.3) obtains angle from any coordinates. MA coordinates (C3,D3) by definition are on the unit circle, whereas MV coordinates (C11,D11) need not be. MV and RV have resultant angle (RA), which is not synonymous with MA although here they are equal because the data are angles or unit vectors.

Macro functions (Sec. 10.2) simplify usage by acknowledging the period k: **kSIN**(k,x) and **kCOS**(k,x) give the proper sine or cosine of x given k; **kANGLE**$(k,$sin,cos$)$ gives angle (translated into system of any k) specified by (sin,cos) pair. **kMA**$(k,$data$)$ obtains MA directly from angles (B4:B10).

	A	B	C	D
		Direction (°)	sin	cos
1				
2	arithmetic mean, *WRONG* ⇒	~~194.29°~~		
3	Mean angle: kMA(360,B4:B10)	**349.44°**	-0.183	0.983
4	data	297.07°	-0.890	0.455
5	data	284.97°	-0.966	0.258
6	data	345.92°	-0.243	0.970
7	data	14.41°	0.249	0.969
8	data	38.24°	0.619	0.785
9	data	42.63°	0.677	0.736
10	data	336.77°	-0.394	0.919
11	averages are coordinates of Mean Vector		-0.136	0.727
12	kANGLE(360,avgsin,avgcos) gives Resultant Angle	**349.44°**		

Two kinds of 'average' angle are available: the resultant angle (RA, average angle weighted by L) and the mean angle (MA, un-weighted). Despite that use of the arithmetic mean of angles is usually wrong, it or the sum may have limited use, e.g. for the amount of torsion tolerated by structures before failure. The issue at hand should determine which measure is appropriate.

4.1.1. CALCULATING BY COMPONENTS (SINE, COSINE)

Key summaries are: resultant vector (RV) and mean vector (MV); their vector lengths RVL and MVL; the resultant angle (RA) of either; and the mean angle (MA, not necessarily equal to RA).

Directions cannot be averaged by summing and dividing by *n*. Because a cycle exists in two dimensions, one-dimensional averages are improper and summaries need to use the sine and cosine components.

Angles can be represented as unit vectors, i.e. with length = 1.0. (Sec. 3.1.2).

A *vector* is either a directed line segment, a direction (angle) with magnitude (length), or a pair of sine and cosine components. It is thus at least 2 dimensional[c]. Its angle can represent a direction or a time in a cycle, and its length can represent some other property (e.g. wind speed). A unit vector is one having length=1.0 (compare Figure 4-1 and Figure 4-2).

(See also Batschelet (1981) for basic conventions (his chapter 12, Essentials of Plane Geometry) and (his section 12.4) algebraic relations and useful trigonometric identities.)

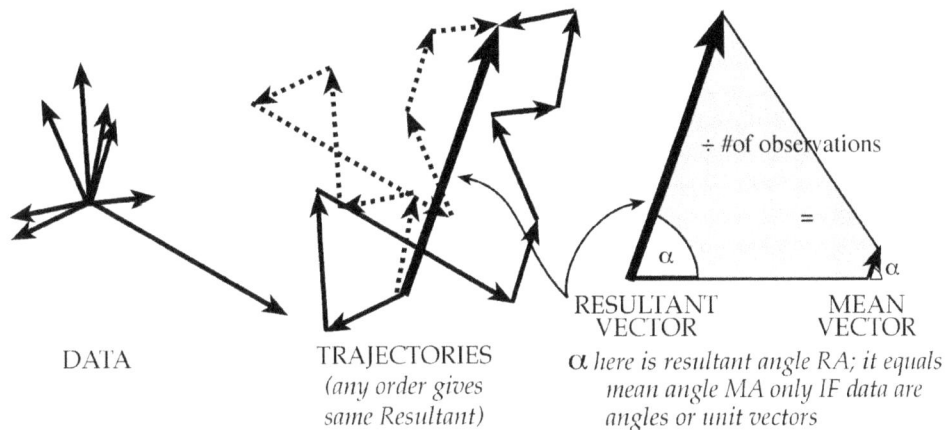

DATA   TRAJECTORIES *(any order gives same Resultant)*   RESULTANT VECTOR   MEAN VECTOR

÷ #of observations

α *here is resultant angle RA; it equals mean angle MA only IF data are angles or unit vectors*

Figure 4-1. Non-unit vector data, resultant vector (RV), mean vector (MV). Data: *n* vectors *i* with direction α and length *L* (i.e. they are not unit vectors) representing, e.g., the movement of a snail in successive minutes. The trajectory is the vectors in sequence. The RV is net total movement; sequence has no effect. RV coordinates are ($\sum L_i * \sin \alpha_i$, $\sum L_i * \cos \alpha_i$); divide by *n* for MV coordinates ($\sum L_i * \sin \alpha_i / n$, $\sum L_i * \cos \alpha_i / n$). Angle of both RV and MV is resultant angle (RA). The data vary in *L*, so at least one is not a unit vector; if all data were angles alone or unit vectors then RV and MV would be unitary as in Figure 4-2, and only then would RA equate to the mean angle MA.

Vectors can be expressed as sine and cosine components, and these are key in calculation of the resultant vector, mean vector, and mean angle (or circular average). We can describe vectors in an (*x,y*) coordinate system; we can write (x'',y'') to avoid confusion with a dependent variable *y* (Sec. 13.3). An vector *i* with angle (α_i) and length (L_i), beginning at the origin, has sine and cosine components:

---

[c] Recalling that a cycle is 2 dimensional, and therefore so is an angle, you could think of a vector as 3-dimensional.

Eq. 4-1: $sine\_component_i = x''_i = L_i \sin\alpha_i$

Eq. 4-2: $cosine\_component_i = y''_i = L_i \cos\alpha_i$

and these give its terminal coordinates $(x''_i, y''_i) = (L_i \sin\alpha_i, L_i \cos\alpha_i)$, i.e. they express the vector in trigonometric format (Sec. 3.4). Sec. 4.3 shows how to convert a (sin,cos) coordinate back to an angle.

A vector's coordinates (x'', y'') equate to (sin,cos) in the azimuthal* visualisation, or (cos,sin) in the polar angle visualisation. (* The default in this book; see Sec. 3.2.)

CAUTION, *judgement required*: depending on the context, if the vector is very small the angle may be meaningless.

Resultant Vector (RV) and Resultant Angle (RA)

Resultant vector (RV): the net distance and direction of *n* vectors. RV has length RVL and angle Resultant Angle (RA), as does Mean Vector (MV), but it does not always equal the Mean Angle (MA). RA can be directly calculated from any coordinates along MV or RV, according to Sec. 4.3.

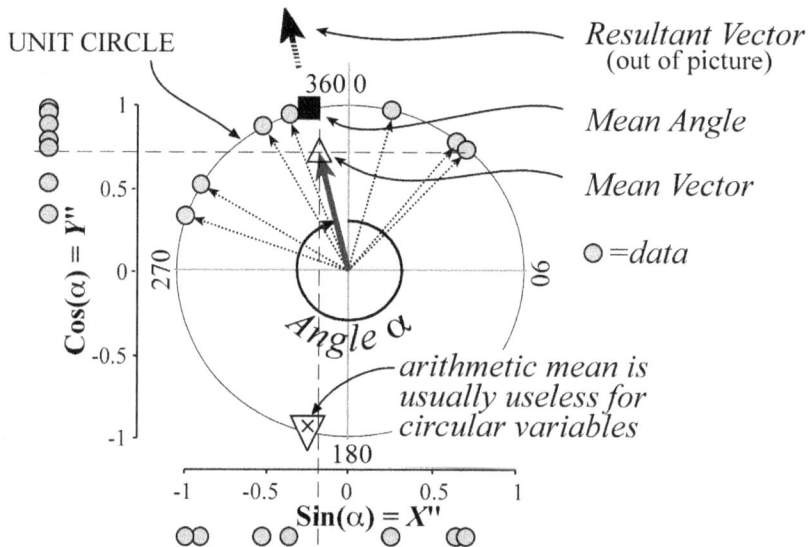

Figure 4-2. Unit data, resultant vector, mean vector, and mean angle. The data (Table 4-I) are angles; angles, by definition, have coordinates (sin,cos) lying on the unit circle, therefore angles can be represented by unit vectors. Summary vectors based on unit data are here called unitary (*cf.* Figure 4-1). The **Resultant Vector** $(\sum L_i \sin\alpha_i, \sum L_i \cos\alpha_i)$ and **Mean Vector** $((\sum L_i \sin\alpha_i)/n, (\sum L_i \cos\alpha_i)/n)$ have direction resultant angle (RA, here α). MV has length RVL$/n$. **Mean angle** (MA) equals the RA only when the latter is calculated from *unit* vectors or equal-length vectors (as here, $L_i = 1.0$, so L has no effect on RA). Arithmetic means or sums of angles would have only rare usefulness, for instance in analysing rotations rather than directions.

The resultant vector is the sum of the 2-dimensional quantities in *n* vectors. E.g., if vectors represented a distance and direction of motion in each of five days, the resultant vector would be a coordinate pair expressing the object's final position after the five days.

A resultant or mean vector is called *unitary* if derived from unit ($L=1$) or equal-length vectors. Nevertheless, and in contrast with the mean angle (MA), the unitary resultant or unitary mean vector itself usually is non-unit (Figure 4-2): $L_{RV}>1$ and $L_{MV}<1$, except in special cases.

The resultant vector is calculated by summing (in any order) each component vector's sine and cosine components separately. The terminal coordinates $(x'',y'')_{RV}$ of the resultant vector (both direction and magnitude) are given by:

Eq. 4-3: $(x'',y'')_{RV}$ = (sum(sine_components),

sum(cosine_components)) = $(\sum L_i \sin\alpha_i, \sum L_i \cos\alpha_i)$

For unit vectors, L_i is 1.0 and can be omitted from Eq. 4-3. The length RVL of RV is by Pythagoras's theorem from its terminal coordinates:

Eq. 4-4: RVL = $(x''_{RV}{}^2 + y''_{RV}{}^2)^{0.5}$

The resultant angle RA can be obtained (see Sec. 4.3 for QC and further detail) from the RV coordinates $(x'',y'')_{RV}$ by:

Eq. 4-5: RA = $\arctan(x''/y'')_{RV}$ + QC

The resultant vector's angle (Figure 4-2) is the same as for the mean vector. You could liken RA to a mean angle weighted by the component vector lengths. If RV or MV is unitary, RA equates to Mean Angle; otherwise, it doesn't.

Mean Vector (MV)

Mean vector (MV): the mean net distance and direction of n vectors. MV has length MVL (=RVL/n). Its angle equals the Resultant Angle (RA), but it does not always equal the mean angle. RA can be directly calculated from any coordinates along the length of MV or RV, according to Sec. 4.3.

The mean vector is the average of the 2-dimensional quantities in n vectors. E.g., if vectors represented a distance and direction of motion in each of five days, the mean vector would be a coordinate pair expressing the object's [proper] average direction and distance per day.

A *unitary* Mean Vector is one calculated from n unit vectors (i.e. length=1.0). Despite deriving from unit constituents, a unitary resultant or mean vector may have length $\neq$ 1.0 (Figure 4-2).

The mean vector (Figure 4-2) has angle RA and length $1/n$ times the resultant vector. Its terminal coordinates $(x'',y'')_{MV}$ are indicated by the point (mean[$L\sin`x$], mean[$L\cos`x$]):

Eq. 4-6: $(x'',y'')_{MV}$ = (sum(sine_components)/n,

sum(cosine_components)/n)

which you may prefer to see written as:

$(x'',y'')_{MV} = ((1/n)\sum L_i \sin\alpha_i , (1/n)\sum L_i \cos\alpha_i)$, or

$(x'',y'')_{MV} = (x''_{RV}/n, y''_{RV}/n)$

Like RVL, the length MVL of MV is by Pythagoras's theorem:

Eq. 4-7: $MVL = (x''_{MV}{}^2 + y''_{MV}{}^2)^{0.5} = RVL/n$

The angle of MV equates to the angle RA of RV (Eq. 4-5). If MV is unitary, RA equates to Mean Angle.

MV finds statistical application in the Rayleigh test (Sec. 7.1).

Mean Angle (MA) or Circular Average

Mean angle (MA, also called *circular average* or *mean direction*): the average of
n directions, angles, or unit vectors. MA can equal RA only if RV (or MV) is
unitary, i.e. calculated from unit vectors.

The circular average or mean angle is the proper average of n angles or unit vectors. It does not, however, equate (Table 4-I, Figure 4-2) to the arithmetic mean of the angles $((\sum\alpha_i)/n)$. Neither is MA always the angle (RA) of the MV, even though that is suggested by the conventional names mean angle and mean vector; MA equates to RA only if the MV or RV is unitary.

MA is a direction or angle with no [vector] length; therefore, its (sin,cos) coordinates are on the unit circle and it equates to a unit vector.

From *n* vectors, obtain MA as follows: first reduce lengths of all vectors to 1.0 or to equal lengths, or ignore all lengths, or correct their (sin,cos) coordinates to the unit circle by dividing by their lengths. Then calculate RV or MV (Eq. 4-3 or Eq. 4-6), and then obtain its angle (by Sec. 4.3.2).

From *n* angles, obtain MA as (mean(sin),mean(cos) as follows:

Eq. 4-8: $mean(sine\_components) = (1/n)(\sum \sin\alpha_i)$

Eq. 4-9: $mean(cosine\_components) = (1/n)(\sum \cos\alpha_i)$

These give the coordinates—on the unit circle—of the mean angle:

Eq. 4-10: $(x',y')_{MA} = ((1/n)(\sum \sin\alpha_i),(1/n)(\sum \cos\alpha_i)) =$
                  $(mean(sine\_components),mean(cosine\_components))$

Notice that, unlike RV or MV, MA does not use any lengths.

From here (see Sec. 4.3.2 for QC and further detail), obtain the angle α from its coordinates $(x',y')_{MA}$:

Eq. 4-11: $MA = \arctan(x'/y'')_{MA} + QC$

The choice of MA, RA, or even a sum or arithmetic mean of angles, depends on the phenomenon being described.

4.2 DIRECTIONAL COMPONENTS OF COMPASS VARIABLES

We often need to consider only a single component of a motion or force. We do the reverse of obtaining a resultant vector. We represent the motion as a vector, with its magnitude and length. We can then easily take the standard components represented by sine and cosine. A more complex but manageable issue would be, for instance, if we want the along-shore and on/offshore components of some measured motion. We could do that by rotating the data (Sec. 4.4, and see Batschelet (1981)) so that our coordinate system is aligned with the shoreline. In the rotated system, the on/offshore and upshore/downshore components would be substantially well represented by either the sine or cosine (depending on how we rotated it). These are just some of the possible applications of directional vectors.

Directions, even with velocity, can be decomposed into two components, sine and cosine (Figure 1-3). E.g., for a fish swimming at 1.38m s^{-1} toward the south-southeast or 157.5°, the west-to-east component C_{WE} and the south-to-north component C_{SN} and are obtained by sine and cosine, respectively:

C_{WE} = sin157.5° * 1.38m s^{-1} = 0.5281m s^{-1} , indicating (it is positive) net eastward motion, and

C_{SN} = cos157.5° * 1.38m s^{-1} = -1.2749m s^{-1} , indicating (it is negative) net southward motion.

Inspecting these together, the net motion is somewhere in the southeast, but nearer south because the magnitude is greater. To get the direction exactly, see Sec. 4.3.

4.2.1. WIND: DEALING WITH STANDARD 'FROM' CONVENTION

This section follows from the sections on vectors. The oddity is that wind directions are traditionally reported (Figure 4-3) as the direction they are blowing *from*, so this has to be acknowledged, and adjusted if needed, in any analysis.

There are purposes for which working in the "from" convention is inconvenient. If so, label data carefully and convert it to the "to" convention in the early stages of working with it. Conversion can be done by several methods, e.g.:

[i] add 180 and take modulo(360) of the result (e.g.: for a wind "from" 215°, add 180° = 395°, modulo(360) of 395° = 35°, because 395/360 is 1 remainder 35); or

[ii] after taking sine and cosine: multiply both sine and cosine by -1 to give the coordinates of the angle in the "to" convention, then (unless the sin and cos are all you need) solve (Sec. 4.3) for the angle that represents.

There are two axes here: the north-south and the east-west. It's convenient to refer to them by the directions that are on the positive ends of the map, i.e. northerly and easterly, and negative values indicate southerly and westerly. For example, given a 23 knot east-northeaster (i.e. wind *from* 67.5°), the easterly component is -sin(67.5°) and taking velocity into account the easterly or horizontal component is -0.9238*23knots, giving -21.24 knots *toward* the positive-X direction; the positive X direction is east, so the negative value indicates motion to the left, or the west. Similarly, the northerly component is -cos(67.5°)*23knots = -8.8017knots in the [negative Y, negative north] southerly direction.

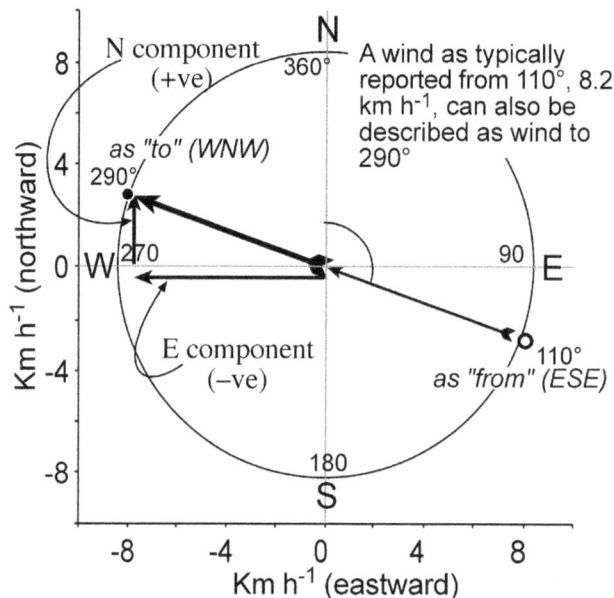

Figure 4-3. 'To and from' and conventions for winds. Arranging the east/west components on the *x*-axis and the north-south components on the *y*-axis. A "from" southeast movement is equivalent to a "to" northwest movement. Knots east- or north-ward are obtained by subtracting the vector's origin from its terminus (the point). In the "to" mode, knots are readable straight from the terminus.

E.g., assuming an 8.2 km h^{-1} wind: northward component is positive: Calculate from "from" as: =−1*cos(110deg)*8.2km h^{-1}= 2.8 km h^{-1}.. or calculate from "to" as:= cos(290°)*8.2km h^{-1}= 2.8 km h^{-1} (the southward component would be the negative of that). Eastward component is negative: calculate from "from" as: = -1*sin(110°)*8.2km h^{-1}= -7.7 km h^{-1}

Method [ii] is obviously the most efficient if we need, anyway, to separate the components for our analysis. Converted sines and cosines then can be multiplied by vector length and mapped on a coordinate grid to portray direction of motion rather than origin. Where a "from" wind vector has length L_i and angle α_i, the graphed points (and directional components) for each "to" wind vector$_i$ are $X_i = L_i(-\sin\alpha_i)$ and $Y_i = L_i(-\cos\alpha_i)$. The length L might represent wind speed, current speed, etc., and its units (e.g. knots, km h^{-1}, m·s^{-1}) appear on the graph axes.

Bear in mind the purpose, and choose the appropriate kind of average. For example:

[a] to estimate wind load, wind-related evaporation, etc., a simple average of wind speed may be useful and the direction component may be of less interest.

[b] For an average direction without regard to wind speed, we need the mean angle (Sec. 4.1.1); i.e. we disregard vector length (speed) or assume all are equal.

[c] For the average or net transport over time (mean or resultant vectors, Sec. 4.1.1) of an imaginary particle suspended in the flow (i.e. the average movement of the fluid), we need data in the form of vectors (both direction and speed).

For the exact trajectory of a drifting particle, the data must be ordered according to times of observation, but for the net transport or the mean vector the order does not matter (Figure 4-1).

4.3   Find angle from coordinates (sin,cos) or (x",y")

This is probably the handiest little trick in the book.

> IN SUMMARY: the true α can be obtained from as $\alpha = \alpha' + QC$, where QC is a quadrant correction based (Figure 4-5, Table 4-II, Table 4-III) on the signs of the coordinates. The uncorrected estimate $\alpha' = \tan^{-1}(x_i/y_i)$, where (x_i/y_i) represent the angle on the unit circle as (sin, cos) or represent the terminus of a vector in $(x",y")$ or (sin, cos) coordinate space. The correction QC is: 0 cycles if 1st quadrant, (x_i/y_i) or (sin,cos)=(+,+); 0.5 cycles if 2nd or 3rd quadrant, cos or y_i <0; 1.0 cycles if 4th quadrant, (x_i/y_i) or (sin,cos)=(-,+). (Refresher note: arctan means the same as $\tan^{-1}$.)

So far so good—we have found that we can manipulate directional data with little difficulty in the coordinate system of sines and cosines.

But how do we translate a pair of coordinates back into an angle? We find that is not as easy as taking the inverse on our calculator. It is tricky. Here's the problem: suppose we have an angle 200°; take the sine and cosine, and see if arcsin or arccos get your angle back:

sin(200°)= -0.34, but $\sin^{-1}$(-0.34) = -20° (not correct)

cos(200°)= -0.94, but $\cos^{-1}$(-0.94) = 160° (not correct)

That value of 200 is, seemingly, lost forever. However, the solution to "what is the angle having *both* sin= -0.34 and cos= -0.94?" etc. is not ambiguous, because although the sine and cosine inverses have two solutions each, there is only one solution common to both. So it is not impossible, merely tedious. We will see below how to deal with it efficiently.

4.3.1. WHY IT'S TRICKY: AMBIGUITY OF INVERSE TRIGONOMETRIC FUNCTIONS

Just as $\sqrt{4}$ ambiguously equals -2 or +2, arcsin and arccos ($\sin^{-1}$ and $\cos^{-1}$) functions have two solutions (Figure 3-3). As with $\sqrt{n}$, the calculator typically gives you only one of the possible two, so it does not alert you that you may have a problem! This means we cannot reliably recover an angle from its sine or its cosine alone. We need both, and that makes it complex. You could plot sine and cosine and then get out a protractor ... too cumbersome of course but at least the thought experiment tells you that the problem can be solved.

For example: sin(240°) = –0.866025..., arcsin(–0.866025) = –60°. What happened? ("–60°" means 360°+(–60)°= 300°—*but that is still not our angle.*) But the other legitimate angular solution exists also: 180-(-60) = 180+60 =240. What did I do? Look again at Figure 3-3, or sketch a unit circle on your table napkin. See that there are *two* angles that could generate the same sine; they are represented by the *two* intersections of the unit circle by the vertical line sine = x = –0.866025.

The "–60°" then also implies the angle on the negative x, leftward, side of the vertical diameter of the unit circle, not just from the top (360° or 1.0 cycle) but also from the bottom (180° or 0.5 cycle). That is the ambiguity, and if you have only a sine (or only a cosine) value, you cannot reduce the number of angular solutions to fewer than 2.

Each sine (likewise each cosine) has two possible angles because all solutions for sin=x'', or cos=y'', form straight lines (Figure 3-3), and any line (e.g. sin=x''=0.707, a vertical) intersecting a circle must do so at two points. (For the sharpies who holler "tangent! tangent!", nice try but we stick to our story and tell the jury that tangency is not intersection).

Thus, in calculators and computers the functions arcsin or arccos (sin^{-1} or cos^{-1}) each return only one of the two possible values (call it α-uncorrected, α', α-prime), and the other is implied: 180°-α' for arcsin, and 360°-α' for arccos.

Thus, the pairs $\alpha'_{returned}$ and $\alpha'_{implied}$ are:

	RETURNED	IMPLIED
from SINE	arcsin(α)	(180° *or* 0.5 cycle)–arcsin(α);
from COSINE	arccos(α)	(360° *or* 1.0 cycle)–arcsin(α).

So, while sin(α) has only one solution per α, arcsin(i) has two possible solutions per i. Note that the calculator only gives you one; as with square roots where the positive one is given, with arcsin the angle between 0° and 90° is given. Only by using both parts of a pair (sin,cos) can a single solution be found for the angle.

Zar in his much-appreciated book (Zar 1984) shows high regard for the reader's resourcefulness when he says, merely, "the angle with this cosine and sine is...". It's true, but it's cumbersome because you still have to trudge through finding the second possible angle in each pair, and match them up.

This may all be quite frustrating, but please don't blame me, I didn't invent trigonometry. In fact, please don't blame trigonometry either, or the beautiful decomposition of angles into rectilinear components (sine and cosine). In fact, the angle is most fundamentally expressed by its (sin,cos) coordinates. And, to be fair, the angle is really two pieces of information: if I tell you the angle is 30, but I don't specify units (degrees, rads, ...), you can't tell what the angle really is—thus, 30° really means an angle of 30/360ths of one cycle.

Fortunately, there is a simple trick to finding α (alpha). It's based on the tangent, but requires a correction for angles that are not in Quadrant I.

4.3.2. TRICK: QUADRANT CORRECTION TO FIND TRUE ANGLE

(Finally ...) we can use a quadrant correction (QC) to get us the true angle α from coordinates (x'',y'') or (sin,cos), or to get ∂ (the phase angle) from the coefficients for the sine and cosine components in a periodic regression.

Suppose (Figure 4-4) we have a vector from (0,0) to coordinates (1.969, 0.347), and we want the corresponding angle α. How do those coordinates relate to an angle? The coordinates can be called (x''_i,y''_i) = ($L_i \sin\alpha_i$, $L_i \cos\alpha_i$) where L is the length of the vector and we want to find α_i. (Did you notice, I made it seem even more tricky? ... the coordinates are well outside the unit circle. That's readily apparent because one of the coordinates is >1.0, so it's impossible for L to be ≤1.0, even with the kindest application of Pythagoras's theorem. That was just to show that the coordinates don't need to be on the unit circle.)

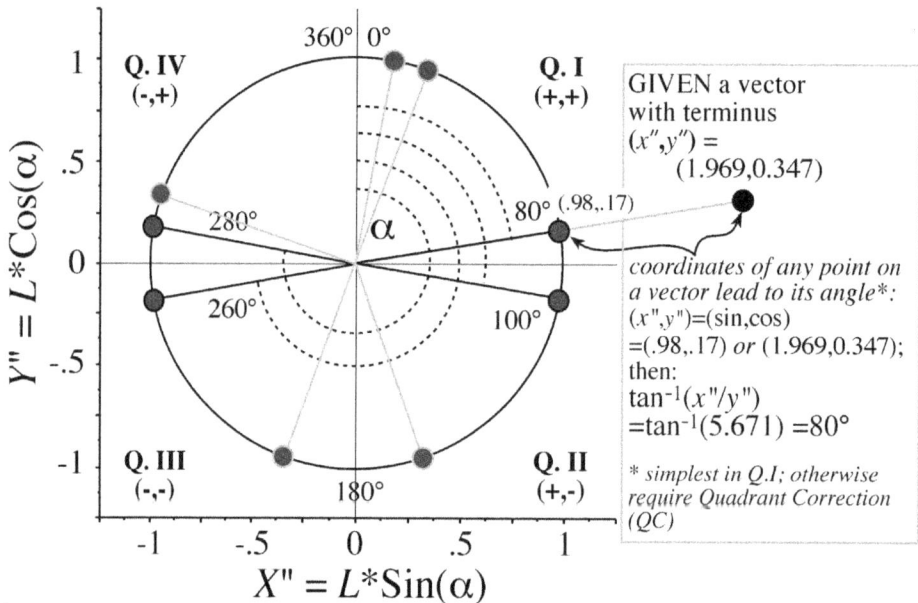

Figure 4-4. Unit circle and vectors with example data from Table 4-II. Unit circle has by def. radius=1=√(x^2+y^2). If a point lies on the unit circle then its coordinates (x'',y'') are equal to the (sin,cos) of the angle, and L of its vector representation is 1. Arctan(x''/y''), with quadrant corrections (QC, Sec. 4.3.2), returns angle from coordinates as in Table 4-II.

For points on the unit circle we *could* (tediously) find α by finding $\alpha'_{returned}$ and $\alpha'_{implied}$ (Sec. 4.3.1) from arcsin and arccos of x'', then choosing the α common to both. Then, points not on the unit circle would require the same, after calculation of the corresponding points on the unit circle by calculating L_i = √($x_i^2+y_i^2$), then multiplying both x_i and y_i

by $1/L_i$ to find (x''_i, y''_i), where x''_i & y''_i would be the sine and cosine of angle α that we want to identify. Tedious! There's a quicker way.

The quicker method (Figure 4-4) eliminates the need to calculate the hypotenuse. It relies on the fact that for a given angle, the ratio x''_i/y''_i for a vector L_i is the same no matter how far L_i is extended; this means we can safely rely on $\tan^{-1}(x''_i/y''_i)$, at least in Quadrant I. Check it: first calculate $\tan\alpha$ as $x''_i/y''_i = 1.969/0.347 = 5.671$; Then we take the arctangent: $\alpha'=\tan^{-1}(5.671...)=80°$. Our example indicates this is correct—so far so good—but[d] we just happened to be lucky with an α that was within the 0-90° interval (quadrant I) which is what the calculator always thinks we are looking for. I.e., *the angle offered by the arctan function is only a part of the information we need to recover the angle.* So, we need quadrant corrections (QCs) to find angles outside the range 0 to 90°.

Table 4-II. Quadrant correction (QC), example angles (Figure 4-4), transformed to sin & cos and then recovered by $\arctan(x''_i/y''_i)$ in azimuthal system, or $\arctan(\sin/\cos)$ in either system. QC rule is given as a fraction of the cycle (or given for $k = 1$). Provisional answers (α') that are initially false are lined through. Azimuthal and polar angle visualisations (Figure 3-2) both identify quadrants and proper QC similarly according to the signs (sin,cos).

EXAMPLES (for angles α in degrees)							RULE		
α deg	$\sin\alpha$	$\cos\alpha$	$\tan\alpha$ =sin/cos	$\tan^{-1}$ = α'	+QC	= α	Quadrant	signs (sin, cos)	ADD QC cycles
	X" or B_1	Y" or B_2	X"/Y" or B_1/B_2	deg | rad					
10	0.174	0.985	.176	10.0	0 | 0	10	I	+,+	0
20	0.342	0.94	.364	20.0	0 | 0	20			
80	0.985	0.174	5.676	80.0	0 | 0	80			
100	0.985	-0.174	-5.665	~~79.9~~	180 | π	100	II	+,−	0.5
160	0.342	-0.94	-0.364	~~19.9~~	180 | π	160			
200	-0.342	-0.94	0.364	~~20.0~~	180 | π	200	III	−,−	0.5
260	-0.985	-0.173	5.687	~~80.0~~	180 | π	260			
285	-0.966	0.259	-3.725	~~75.0~~	360 | 2π	285	IV	−,+	1.0
290	-0.94	0.342	-2.743	~~70.0~~	360 | 2π	290			

The signs of sine and cosine tell us what quadrant we are in. Then we can add, to $\arctan(\sin/\cos)$, a correction that is conditional on what quadrant our angle is in. For Quadrant I, where (sin,cos) or (x'',y'') are (+,+), add 0°. For Q II or III where (sin,cos) are (+,-) or (-,-), add 180°. For Q IV where (sin,cos) are (-,+) add 360°. Of course, equivalent corrections for working in rads are either 0, π or 2π.

---

[d] (*Recall from Figure 3-3 that there are TWO α's which will give each tan ($\tan(260°)$ also = 5.67), and there are still *two more* α's that will give the same magnitude tan but with *different* sign)

Quadrant corrections allow recovery of an angle from its (sin,cos) pair, *except* where cos = 0 (angle is either 90° or 270°). These are special cases (Batschelet 1981), because if cos = 0, then sin/cos = sin/0 (undefined). These however are the easiest to deal with: e.g. the point (0,-24) is on the negative *Y* axis, therefore α=270°; (0,11) is on the positive *Y* axis, therefore α=90°.

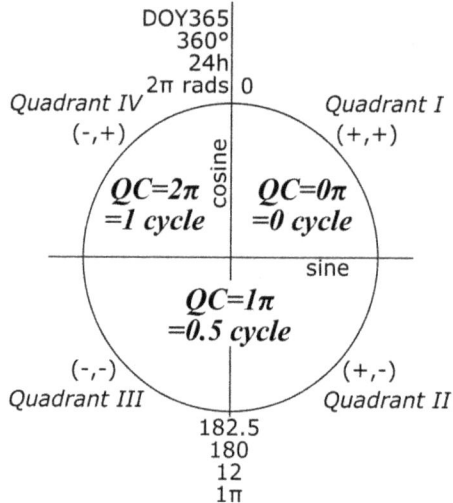

Figure 4-5. Quadrant correction (QC) method for recovering true angles (α) from (sin,cos) pairs; this visualisation is for the azimuthal system. True α = arctan(sin/cos)+QC. See also Table 4-II and Table 4-III.

Table 4-II and Figure 4-5 can be similarly expressed (Fisher 1993) as a conditional calculation, from (x'',y'') or (sinα,cosα), of the true angle α:

Table 4-III. Quadrant correction (QC) method as a conditional equation (Fisher 1993). It can be applied to regression (Sec. 6) coefficients for sine and cosine components to determine phase angle for each cycle. A spreadsheet formula for QC in radians is: "=2*pi()*IF(COS<0,0.5,IF(SIN>=0,0,1))".

	equals	IF	Quadrant (azi. or pol.)	signs (sin,cos)
α=	$\tan^{-1}(\text{Sin}\alpha/\text{Cos}\alpha)$ **+ 0 cycles**	Sinα & Cosα> 0	Q. I	(+,+)
	$\tan^{-1}(\text{Sin}\alpha/\text{Cos}\alpha)$ **+ 0.5 cycle**	Cosα < 0	Q. II, III	(+,-), (-,-)
	$\tan^{-1}(\text{Sin}\alpha/\text{Cos}\alpha)$ **+ 1.0 cycle**	Cosα > 0 & Sinα < 0	Q. IV	(-,+)

4.4   ROTATING CIRCULAR DATA

Oh, sorry, my section title is wrong. Batschelet (1981) points out that rotation of data is in effect a rotation of the reference plane; that is elegant, because the data are factual and the plane is a convention for referring to them. By analogy: the plates, spoons, knives and forks remain in place while the table is rotated.

"Data rotation" therefore means finding new coordinates of the same data on a rotated reference plane. Our plane is usually in $(\sin{`}x,\cos{`}x)$ or $(\sin{`}\alpha,\cos{`}\alpha)$ space—which we also call (x'',y'') space (Sec. 13.3). You can in principle rotate any (x,y) including non-directional data, but it may not always make sense if, e.g., (x,y) is (age,height).

Conceptually, data rotation is simple. Imagine a graph (a polar plot) showing a number of measurements of wind direction and speed. Suppose you need to rotate those data, just as if you rotated the printout of the graph to look at it from another angle. This is perfectly legitimate with directional or angular data because the zero of a cycle is arbitrary.

Why might you want to rotate (e.g.) wind data? Perhaps your data come from a situation that has its own natural axis, perhaps a shoreline or a reef, or perhaps you are not a biologist at all but an engineer considering the best orientation of a linear structure like a wall. Perhaps (biologist) you have data from several shorelines that are not parallel, and you want to express all observations in terms of shoreline orientation so that you can put them together. There are many possible reasons to want to rotate data.

Vectors can expressed as [i] angles α (the circular X) and lengths L, where L can be any Y variable; or as [ii] coordinates (x'',y'') which integrate both angle and length as $(L \sin{`}\alpha, L \cos{`}\alpha)$ or $(L \sin{`}x, L \cos{`}x)$. We can rotate angles, or rotate coordinates.

Rotating angular data is simple: add a rotation angle to the data angles; then, for any results that are either ≤ 0 or $\geq k$, correct[†] to $0 \leq$ result $\leq k$. Rotation doesn't affect vector lengths.

> † Recall that (sin,cos) are the fundamental descriptors of an angle, and $\sin{`}x=\sin{`}(x+k)$, so adding or subtracting k (1.0 cycle) results in the same angle. Therefore, correction of the result is possible by: [a] adding or subtracting k until the result is a positive value ≥ 0 and $\leq k$; or [b] dividing a positive result $\geq k$ by k and keeping the remainder or adding nk to a negative result and then doing the same; or [c] taking proper sine and cosine and then [re]obtaining the angle as in Sec. 4.3.2.

Thus, a crude way to rotate (x'',y'') data would be to re-calculate all vector lengths (using Pythagoras' theorem: $L=(x''^2+y''^2)^{0.5}$), and angles (using arctan and quadrant correction, Sec. 4.3.2), then add the rotation to each angle, correct results to between 0 and 1.0 cycles, then get the results into coordinate form again.

More efficiently, Batschelet (1981) shows how to directly rotate (x,y) (by which he means our x'',y'') data—i.e. find their locations on a rotated reference plane. It is easy to imagine your graph of vectors (each running from (0,0) to (x''_i,y''_i)) and imagine swinging all vectors by the same angle around the origin.

Rotation of one bivariate data point (x'',y'') requires two calculations: one each for the rotated x and y. Each of those calculations uses both the original x and y.

Batschelet uses the polar coordinate visualisation, in which rotation by a positive angle is counterclockwise. In the azimuthal systems a positive rotation is clockwise, so we can simply change the sign of the rotation.

Table 4-IV. Example of original data, and the same data rotated by $\alpha = -\theta = 60°$ or 1.0472 radians; graphed in Figure 4-6.

Original		Rotated by α=60°	
sin	cos	sin	cos
x''	y''	x''	y''
3.2	5.6	6.44	0.03
13.0	24.0	27.28	0.74
3.0	8.0	8.43	1.40
-5.0	-9.0	-10.29	-0.17
2.0	-11.0	-8.53	-7.23
-13.0	18.0	9.09	20.26

I'll give two sets of formulas, one for the Azimuthal visualisation where rotation of (x''_i,y''_i) by a positive angle is clockwise, and one for the Polar Angle visualisation where it is counterclockwise. Where subscripts *orig* and *rot* mean original and rotated data coordinates (x'',y''), clockwise rotation by positive α (alpha; azimuthal consistent) is

Eq. 4-12: $x''_{rot} = x''_{orig}{*}\cos\alpha + y''_{orig}{*}\sin\alpha$

$y''_{rot} = -x''_{orig}{*}\sin\alpha + y''_{orig}{*}\cos\alpha$

and counterclockwise rotation by positive θ, (theta; polar consistent) is:

Eq. 4-13: $x''_{rot} = x''_{orig}{*}\cos\theta - y''_{orig}{*}\sin\theta$

$y''_{rot} = x''_{orig}{*}\sin\theta + y''_{orig}{*}\cos\theta$

Choose the set of formulas that is consistent with your notion of positive rotation, and then control direction of rotation by the sign of your alpha or theta. Changing the sign of the rotation is equivalent to changing the formula.

Why do the two sets of formulas differ only in the signs of the sine components, and the cosine components are the same? Simple: while $\sin(-\alpha) \neq \sin(\alpha)$, $\cos(-\alpha)=\cos(\alpha)$. Changing sign of angle (e.g. 47° and −47°) changes the sine, but has no effect on cosine. Don't believe me? Try it. Why does the cosine's sign remain unchanged? Because (see Figure 3-3) positive and negative angles have *sine* components falling on *opposite* limbs of the *x*-axis, but *cosine* components falling on the *same* limb of the *y*-axis. Cosine = adjacent/hypotenuse, and the adjacent side will be the same for a positive or negative angle, in either the azimuthal or the polar angle visualisation.

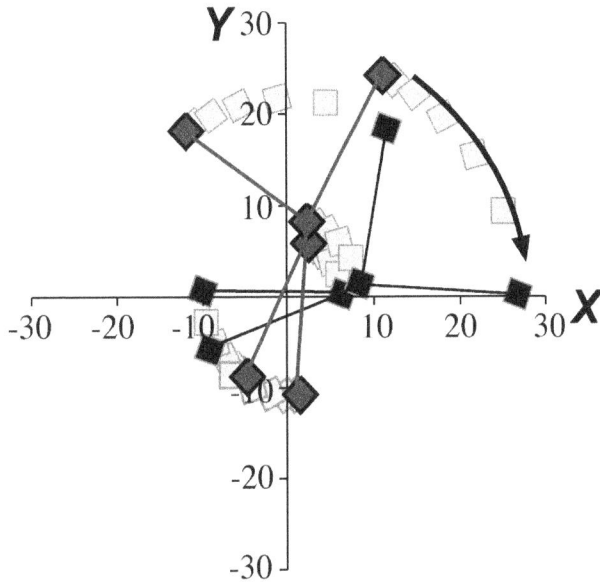

Figure 4-6. Original and rotated data from Table 4-IV.   Rotation is 60° = 1.0472 rads. Points are joined, just to show which set they belong to.

In a spreadsheet therefore you will have a pair of columns for (x'',y'') coordinates of original data, and a pair of columns of formulas to recalculate rotated coordinates. An example is shown in Table 4-IV and Figure 4-6.

As Batschelet (1981) carefully points out, the rotated data will preserve all vector lengths and all angular differences between data points. That's why, when graphics programs use formulas like this to cause Mickey Mouse to tumble, rotated Mickey still looks like original Mickey.

4.5   FINDING THE (smallest) DIFFERENCE BETWEEN TWO ANGLES

Although the difference between two numbers is a simple construction, it's a little tricky with circular variables. We'll may easily visualise the difference between two angles on a graph, but calculating it is less simple. In Figure 4-6, we rotated data by 60° ... or was that by -300°? Either can give the same result. Sometimes we'll want to find the smallest difference between two angles.

Batschelet (1981) "denote[s] the angular distance by the symbol $|\phi, \psi|$ ... $0° \leq |\phi,\psi| \leq 180°$", ϕ and ψ (phi and psi) being the two angles in question, and we'll preserve that notation in this section.

The difference between two angles would seem to be a very simple matter, but Batschelet (p. 242) cautions: "the calculation of the angular distance is not trivial." The key is that, to get to one angle from the other, we could go clockwise or counterclockwise, so there are two angles that could stand for the difference $|\phi,\psi|$. Batschelet of course means the smaller arc (i.e. don't say -10° and +10° differ by 340°) is the angular difference. So the definition (Batschelet 1981) is:

Eq. 4-14: $|\phi, \psi|$ = smaller of the two angles $|\phi - \psi|$ and

$360° - |\phi - \psi|$

That being slightly cumbersome however, Batschelet suggests an elegant shortcut:

Eq. 4-15: $|\phi, \psi| = \arccos [\cos (\phi - \psi)]$

Note that appropriateness of this measure depends on the question. Sometimes the angle of interest is not the smallest one. For example it may be that something rotates many times per second (etc.), and if the process of rotation (rather than the final position) is important (as in the amount of rotation generated by a crocodile to tear pieces off its prey) then we might not want "the smaller of the two angles". Or if you are measuring accommodation of torsion in a flower stem that turns to face the sun, or in a flexible element that has to run through a pivot and survive torsion, you want the actual angle that was turned, whether it is the larger or the smaller of the two possible angles (and there would be examples of this situation in engineering).

5   DATES & NATURAL CYCLES

When we have observations indexed to time, and need to analyse in terms of time, we need (a) well-recorded times of the observations, and (b) to get the time data into a form that can be analysed.

Everything here is simple, but if you are not organised it can be tedious and can generate spectacular errors. Automated conversions (Sec. 10) in spreadsheets smooth workflow and reduce the opportunity for error.

5.1   DATE AND TIME: RECORDING AND FORMAT

Conventions can be arithmetically proper, or not, for the numbering of months in a year, or days in a month. I'll use the phrase "*common format*" to denote the socially accepted systems for measuring time; these are complex, not to say awkward. What the decimal system did for currency and distance has not yet been done for time. Larger units contain smaller units in sequence numbering 365 days (or 12 months that in turn contain either 28, 29, 30 or 31 units) then 24, then 60, then 60 again. These irregular and inconstant divisions don't match our decimal counting system, so dates and times are generally useless for graphing or analysis until transformed.

While we have successfully adopted the zero in algebra, zeros are absent in common formats for time (the abominable format 1/1/2001 indicates, commonly, the first of January). For calculation and graphing, dates need conversion to something arithmetically proper. We need to express a time coordinate as (e.g.) month 0 (for Jan 0) or 0.5 (for mid-Jan), or as day 0, etc.; we do this using the day-of-year (DOY, 0-365) format. In converting results from *true* (such as DOY) back to any of the *common* formats, you have to reverse the transformations.

It will help if you adopt a mental concept of each cycle as a range 0 to 1; any fraction of that can be multiplied by the cycle's k to get days in the year, hours in the day, etc.

5.1.1.   MONTH (COMMON FORMAT) VS. MOY (MONTH-OF-YEAR)

Months are commonly numbered 1-12 (common) instead of 0-11 (proper). In cases where date is to be analysed using the month number as a measure of the time in the year (season), remember that common format (i.e. Jan as month 01) lacks a zero. Therefore the *proper* month (MOY) is the *common* month less 1. When calculating peaks from regressions done using *proper* month, the peak will also be in *proper* format (whether in radians, degrees, MOY, DOY, etc.), and if it is needed in *common* format then be mindful of the (de-)correction required. If you calculate using common format month numbers, you will get an incorrect result with no clear-cut correction possible; you'll have to do it all over or just be wrong.

It would usually be better to do your calculations using day of year, see Sec. 5.2: DOY (Day-of-Year), and DOY table Sec. 16.

5.1.2. IN FIELD NOTES: YYYYMMDD.hhmm

When sampling: it is bad discipline to ignore data, so always record all dates with times of day, as time of day often affects either phenomena or observation or both.

Field notes should accommodate our habits (and some common formats) but not at the expense of clarity. Formats like 08/03/97 are dangerously ambiguous (in some countries they mean month/day/yr but in others they mean day/month/yr) and so should be avoided at all costs.

Common-format date & time can be written conveniently, clearly and concisely as a "field-format date", a composite number of format "YearMonthDay.TimeofDay" as "YYYYMMDD.hhmm" e.g. 19980423.1547 for 47 minutes past 3 p.m. on the 23rd of April 1998. This format is easy to sort automatically by date because, unlike the other common formats, it preserves a left-to-right descending order of units. It allows quick sorting and it's safer because Excel will then treat it as a number (not a date subject to preferences settings of different 'start dates'). It's also handy for field notes, and though is a "common" format (month 1-12 instead of month 0-11) it is easily converted to proper format (DOY, see table Sec. 16) when needed. Avoid symbolic separators because those cause some spreadsheets to "reinterpret" the date. It is easy, once you have YYYYMMDD.hhmm, to write formulas that retrieve any component (Sec. 10).

The ISO-8601* standard recommends a format for recording common-format date & time. It is virtually the same as the standard above except for my decimal mark; ISO 8601 offers the use of a capital T to separate data and time portions (e.g. 19980328T1955), or no separator at all, where I use a decimal. The T would have the advantage that a spreadsheet would not attempt to convert it to a clever date format. The disadvantage would be that we could no longer use arithmetical operators (e.g. multiplication, division, and the Integer function) to dissect out the components, but we could still use string-handling formulas (e.g. the operators LEFT, RIGHT, MID, and SEARCH) provided the spreadsheet would accept the results as numeric and available for calculation.

* www.iso.org/iso/support/faqs/faqs_widely_used_standards/widely_used_s
tandards_other/date_and_time_format.htm

Alternatively, store common-format YYYY, MM, DD, hh, mm in separate columns. This is a bit more tedious for data entry, and doesn't save any time overall because it's easy to make formulas to process your YYYYMMDD.hhmm format data into columns, DOY, etc.

Record the time of day! When recording data, please don't skip recording time of day (hhmm) just because you don't think it matters. If you didn't write it down, you can't go back and get the information. Not only can time of day can influence biology, it can influence your observation of a system.

5.1.3. IN DATA FILES

Annotate and archive: Incredibly, some people actually throw data away. Some just lose it easily. That has a comical or pathetic aspect, but it also has an ethical dimension.

There is nearly always some component of your work that relied on an external subsidy, not least by the organisms you worked on and your responsibility to science; that brings on you an obligation to do your best to make the data worth keeping, and then ensure that they are kept in a way that supports the possibility of their future usefulness. If it was worth publishing, or worth doing, the data should be well archived.

> **CAUTION: convenience date formats** (e.g. 09/04/97) **in spreadsheets can corrupt data**. *Dates in those formats can spontaneously change simply due to moving the file to a different computer, or re-setting the program's preferences, or doing an upgrade, etc. Therefore, if you have already got those formats, at least add a column with some reference dates from your notes in quoted form to let you check accuracy and correct the preferences. If your preferences changed or you switched computers or reinstalled the program in the middle of entering data in 'spreadsheet convenience' formats, you will have a lovely mess.*
>
> Why does that happen? Excel, for example, allows (set in the preferences panels) several different dates to be chosen as the program's "zero" dates: the date displayed is an integer number that is converted to $x/x/x$ proprietary format by adding some value to the "zero" date; that's why moving your file to a system where the preferences have been set differently can CHANGE your dates by so many years, months and days (effectively destroying your data). Always state what your system is, and annotate your data files well: preserve enough dates in text-only format to allow a quick check of accuracy. Excel hint: if you want to be sure Excel leaves your dates untouched, select "number" format; alternatively, force text format by preceding them with a single or double quote, or use format YYYYMMDD.hhmm.

Therefore, annotate your field data so that someone else can figure out what you did and what it means (if someone else can, then you will be able to also). Include photos and drawings to help document what you did, and how and why. If you do that, then you yourself will be able to do the same thing a few years later when you find another question that could be addressed by the data. Archive in multiple locations electronically, and also keep your original paper notes safe. Ask your department for its archiving service; some national granting agencies now require data to be kept available for at least 10 years.

If you are writing a thesis, and if the data are printable in less than a few dozen pages, consider including them with your thesis. If you want to, you can include a note that use of the data requires your permission.

Publish precise data: Publications often give data tables showing month only (no date). This is regrettable because it throws away data and assumes a step-like transition between months. Or did they forget what day they took the data? ... or not know how to calculate DOY for a better presentation?

5.2   GETTING MULTIPLE CYCLES FROM ONE INDEX *X*

If *Y* is measured at time *X*, then we can express any cycle from *X* many ways (decimal year, day of year, even minute in year). Not only that, but one index *X* can have meaning on multiple cycles (Sec. 5.3, Figure 5-1). Natural cycles of most interest to biologists are the year (seasonal), the lunar month, the day, and the tide. These arise from the Earth's orbit about the Sun, the lunar orbit about the Earth, the Earth's rotation with respect to the Sun's direction, with the tide being the effect of many components. All can be obtained from a single time index *X*. Many cycles show a statistically significant second harmonic (Sec. 13.4).

For example, DOY if sufficiently precise also provides the daily (or lunar, or tidal) cycle. E.g. (Table 5-I) if DOY=128.32, the date is May 9th and the decimal time of day is 0.32, i.e. 0.32 of a day past 0000h.

Table 5-I. Multiple cycles converted from one index *X*. There are many ways of obtaining sine and cosine on multiple cycles from a single linear time datum.

Get sin & cos for finer cycles from same data format, e.g. *X*=**May 09, 0741h**					
TO OBTAIN *sin & cos*: for time of:	from *X* as:	multiply by R`=2π/*k*	to get: R`*X*	& take sin, cos sin`X` cos`X`	
YEAR	DOY= 128.32	2π/365= 0.0172	2.209	0.803	-0.596
DAY	DOY= 128.32	2π/1= 6.2831	806.3	0.905	-0.426
DAY	decimal year =DOY/365 = 0.3515	2π/[1/365] =2π*365= 2293.3	806.3	0.905	-0.426
DAY	DOY fraction= 0.3200	2π/1= 6.2831	2.011	0.905	-0.426
lunar month	DOY= 128.32	2π/28.5= 0.2204	28.29	-0.015	-1.000
lunar month	DOY/28.5= 4.5024	2π= 6.2831	28.29	-0.015	-1.000

Conversion is simple (Sec. 3.5) and sines and cosines can be quickly checked against the quadrant (Figure 3-3). E.g., DOY 128.32 is more than 1/4 year and less than 1/2 year, thus it is in the second quadrant, so the sine *must be* positive and the cosine *must be* negative; sure enough. Likewise, 7.68 h is in the second quarter of the day, so (sine,cosine) must have signs (+,-) ... check it ... sure enough.

5.3   NATURAL CYCLES: HOW TO OBTAIN AND TREAT THEM

Time in most natural cycles can be converted from our common measure of time (date)—some directly, some indirectly. The daily cycle (24h) is very easy, directly obtainable because we already recognise time of day in our common measurement of time. Other cycles, e.g. lunar or tidal, are nearly as easy, requiring some look-up in tables.

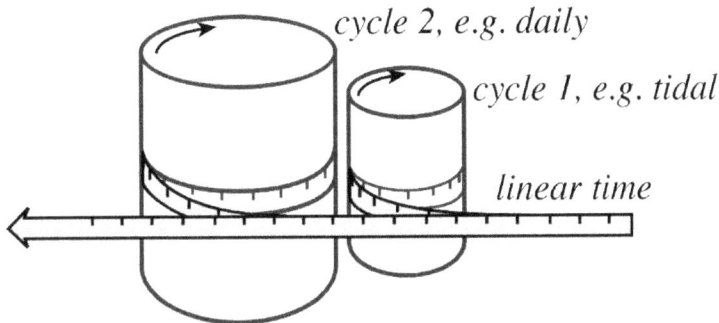

Figure 5-1. Relation of a cycle to linear time. The linear time index can be converted to cycles of any period (circumference), and to multiple cycles at the same time. The analogy is time as a long thread that can be 'wound' on spools representing cycles of any period (circumference), or (similarly) the same linear time can be 'printed' on multiple cycles at the same time.

The use of the same raw X (continuous time index) for more than one cycle may seem counterintuitive, but imagine a tractor, with the little front wheels representing a short cycle, and the large rear wheels representing a longer cycle, rolling over the same road that represents continuous time. Or, think (Figure 5-1) of time as a continuous tape (with limitless copies) that can be wound on cylinders representing cycles with different periods. Continuous time can be "wound" onto a cycle because $\sin(40°) = \sin(360°+40°) = \sin(n(360°)+40°)$, etc. *In other words:* the real cycles keep rolling through time; mathematically, it doesn't matter how many times the tape or thread goes around the cylinder, any point along the string occurs at some angle that represents time or stage in that cycle. The tape can be wound around a cylinders representing any cycle, whether the day, the lunar month, the year, etc. The same point on a string, the same linear time reference, indicates a time on any cycle you care to choose.

This section briefly describes the main natural cycles and suggests reasonable ways to treat them and what period length should be used. A good general habit is to think of position in any cycle as being from 0 to 1 (one complete revolution). Position in a cycle can be identified as "Time Of [Day, Tide, Year, etc.]", abbreviated TOD, TOT, etc. and, if needed, followed by a distinctive letter to indicate the units; for example time of day could be TODd or TOD1 for decimal (0-1), or TODh, m, or s for hours

(0-24), minutes (0-1,440), or seconds (0-86,400), whatever suits your needs.

Virtually all natural cycles (lunar month, tide, year) have irregularities, but most are small and can be ignored. Some shortcuts taken by some authors in respect of cycles are acceptable, and some are not. Bliss (1958) felt it reasonable to treat all months as being of equal length, but that was a computational economy he himself probably would not have advised for those with access to automated computing.

There is no virtue in adding rounding error to data or analyses, so do all analyses with the most accurate values that are reasonable. Don't be hounded by people who exclaim that you only have 2 significant digits! ... the final 3 digits of 4.3278 are no less significant than the final 3 digits of 4.3000, which is what your calculator 'sees' if you give it a rounded number. It is your final result, not intermediate calculations, that should be rounded according to presentation requirements.

(See Sec. 10 for Spreadsheet & programming formulas for data preparation and regression interpretation.)

5.3.1. CYCLES *DIRECTLY* OBTAINABLE FROM DATE

Year:

A year is the time between successive appearances of the Earth at the same point in its orbit around the Sun. The period k is approximately 365 days, a day being the time between successive zeniths of the sun as seen at a particular location on earth. But like most natural cycles, there are complications and irregularities.

More precisely the year is 365.25 d, by which each fourth year is a leap year, but that leaves smaller irregularities: the "leap year" is null if the year is also a century; the century leap year is considered 'on' again if ... i.e. the rule is very complex to maintain accuracy, but that complexity is largely unimportant for most bio-medical issues.

Thus we treat the year simply, and adequately for most purposes, as 365 days. The increase in precision by acknowledging the leap year is rarely worth the bother, as agreed by authorities of the stature of C.I. Bliss who treated the year as 12 months (disregarding differences in month length). Months vary in length, so they are not very nice units.

DOY (Day-of-Year)

A handy DOY table (zero-based, 0-365) is given in the Appendix Sec. 16, and spreadsheet formulas are given in Sec. 10.

The yearly cycle is efficiently represented as Day-of-Year (DOY), where DOY is 0 to 365. DOY=0 (zero) is the first moment on Jan 01 and 364 means the first moment on Dec. 31. 10 a.m. Jan 01 would be DOY 0.41667 (= 0 + 10/24), and one minute before midnight Dec 31 would be 364.9993056. DOY 365.5 = DOY 0.5 of the subsequent year.

It is mathematically improper to number the days 1 to 365, and naive to not recognise the possibility of a decimal fraction of a day; recognising the decimal unavoidably reminds that the year must start at 0, not 1. The DOY scale must be able to accommodate decimals if it is mathematically proper; the only way to do this and not have values over the actual year length (midday Dec. 31 cannot be 365.5) is to have the year begin on day 0; then, including decimals, midday on that first day would be DOY 0.5.

DOY is not habitual to us, and field measurement should accommodate our habits in the interest of accuracy. Therefore, use something like "YYYYMMDD.hhmm" (Sec. 5.1.2) for field notes and durable data records, and DOY for calculation mode. Alternatively, store YYYY, MM, DD, hh, mm in separate columns. *Whatever you do, avoid the 'convenience' formats in spreadsheets (see Sec. 5.1.3).* [*Month numbers "MM": use 'common' format (01-12) for field notes, convert to 'proper' DOY (0-365), then to radians etc. to calculate sines etc.].

Please don't use the term "Julian Day" to mean Day-of-Year; just say Day of Year. To call the DOY the Julian Day has been termed "an illiteracy", though I must confess I once was thus illiterate ... Here, in its remarkably brief entirety, is a short article (Wilimovsky 1990) clarifying the situation:

Wilimovsky, N. J., 1990. Misuses of the term "Julian Day". Trans. Am. Fish. Soc., 119:162.

"The term Julian Day or Julian Date (JD) has an International definition that sets JD=0 as 1 January 4713 years B.C.; see for example, the 1987 "Astronomical Almanac," issued by Her Majesty's Nautical Almanac Office, Royal Greenwich Observatory, on behalf of the Science and Engineering Research Council (her Majesty's Stationary [sic] Office, London). Julian Date is a very precise unit that currently requires at least seven digits (usually more) for its expression. For example, Julian Date 244 7770.5 is the beginning of 1 September 1989. The decimal point highlights another feature of the unit: the Julian Day begins at noon. A number of recent papers that use the term Julian Day are also incorrect in that the authors clearly are referring to a day which begins at midnight.

Recent literature indicates that some investigators are unaware of the accepted convention; they use "Julian Day" to numerically designate days of the 365-d (or 366-d) year, and begin counting days at midnight.

The Julian Day number was devised by Joseph Scaliger. It is named for his father—not for Julius Caesar's calendar."

(Reproduced by kind permission, American Fisheries Society)

What about leap years? Leap years have no biological meaning, they are simply a convenience allowing the calendar to accommodate the fact that the time of Earth's orbit around the sun (our year) is not an integer multiple of the time of Earth's rotation on its own axis (solar day). Their purpose is not to keep each year's calendar equal to the next in astronomical terms, but to prevent small errors from accumulating over centuries. Is it worthwhile to add 1 to each DOY following Feb. 28, and convert DOY to radians by $2\pi/366$ instead of $2\pi/365$? Such a small difference will virtually never be meaningful in an analysis. It is easy enough to make a superficial correction, but which actually gets no closer to the fact, and that is obviously pointless (unless the entire study falls within a single leap year). If there is an observation on Feb. 29, call it Feb 28 (accept an average error of about 0.25/365.25) and carry on, knowing that we are operating far more precisely than we probably need.

Our calendar is a good approximation that maps the path of the Earth in its solar orbit. If we needed greater precision, we would use the Earth's position as an angular measure from an arbitrary zero in its orbit around the Sun. What zero? ... if we used an equinox, we'd need a correction for that because they precess. For Earthbound problems the tiny errors eliminated would rarely make this large trouble worthwhile.

nDOY or sequential day since arbitrary date

For multi-year analyses there's often a need for a term describing the linear march of time. A variable may change not only with seasons or other cycles, but it may also respond to the simple passage of time. For that linear time term, it is useful to pick Jan 01 of the first year of the study as day 0 (DOY=0 of that year and nDOY=0 of the study) and use a continuous count of days (nDOY) elapsed since (DOY 10 of the second year would be nDOY=375). For any day, sin`DOY=sin`nDOY, just as $\sin20° = \sin(360°+20°) = 0.342$; i.e., adding complete cycles to an angle does not affect its sine or cosine at all.

Day (Solar)—Time of Day (TOD):

Period k is the time between successive meridional transits of the sun as seen at a particular location on earth. Conventionally this cycle is divided into 24 hours, or 1440 minutes. It is often called circadian or diurnal. Record time of day in 24-h format (1537h = 3:37p.m.) and convert it to decimal Time of Day ($0\leq TOD\leq1$) for analysis (see Sec. 10).

5.3.2. CYCLES *INDIRECTLY* OBTAINABLE FROM DATE

Indirect relation to date means there is a need for some special calculation, or (e.g. tides) table where the date is an index according to which values for a specified cycle are accessed.

Lunar month and phase:

Period k is the time between successive instances of the same phase of the moon—about 29.5 days. *Note that this has nothing to do with where overhead the moon is seen, but only to the position of the Moon relative to the Earth and the Sun.* The rotation of the Earth and the Earth's meridians are irrelevant. Therefore the lunar phase has reference everywhere at once and you therefore only need one file of lunar phases.

When looking for lunar phase data, you are not restricted to the calendar from the local garage, or MacDonald's Almanac. If you are going to go to the trouble of setting it up in a spreadsheet to relate it to other data via LOOKUP commands (see Sec. 10), you might as well get the good stuff. A good source is the U.S. Naval Office:

http://aa.usno.navy.mil/data/docs/MoonPhase.html

Choose as arbitrary zero (t_0) a well-reported phase like new, first quarter, full, or last quarter. Lunar phases are usually obtainable in UTC (also called GMT) time. Phase is global: if the instance of New Moon is on July 3rd at 0453 UTC, at that moment it is New Moon wherever you are on Earth; relate that instant to local conditions by converting UTC to your local time.

Lunar k is non-uniform, irregular; one lunar month can differ from the next by a day, so it's inadvisable to divide all the dates in your data by 29.5 days and take the fractional portion (inaccuracies will accumulate over the series). The phase of any observation is best obtained by reference to the most recent instance of the chosen arbitrary zero. Supposing you chose new moon (NM) as zero, relate observation times ($t_{observed}$) to lunar phase by looking up the most recent instance of NM as t_0; then, $t_{observed}$-t_0 is a non-standard angle with k=29.5 d approximately or k = the average period for the time range you are looking at. Then, convert that non-standard angle, e.g. to radians (k=2π) as:

$$(t_{observed}\text{-}t_0)*2\pi/29.5\text{d}$$

where t is in days. (See Sec. 3.5.1 for conversions, Sec. 10 for automation.)

Tide:

Period k is the time between successive occurrences of a similar tidal phase (e.g. High to high, low to low). Tidal amplitudes follow a temporal pattern corresponding to the second harmonic (Sec. 13.4) of the lunar phase such that, with a slight lag (called the tidal lag), the greatest tidal amplitudes follow new moon and full moon (similar or opposite alignment of moon and sun). Period is often approximately 12.5 hours, but tide patterns, behaviour and timing are highly location-specific, so look for tide data for the location concerned. To code an observation time into tidal phase, look up (see section on spreadsheet formulas, Sec. 10) the previous time of the arbitrary zero from the best data or predictions (tide tables) and take the difference between that; expressing the difference in the same units as the period, the angular transform to obtain (e.g.) radians is:

$$\text{difference}*2\pi/\text{Period}.$$

Week:

The week is an artificial cycle. It may well have relevance to the activities (and consequences) of humans, but no natural organism should show a weekly rhythm unless it is influenced by human activities. A 7- or 14-day signal could emerge from short tidal series, but a true tidal period would be more defensible. The week, if considered a cycle, is probably not simply sinusoidal, and it may therefore be difficult to obtain a regression with well-behaved residuals unless several harmonics are incorporated. To transform to radians, multiply time of week (day 0-7) by $2\pi/7$.

Day (Sidereal):

Period is the time between successive meridional transits of a given star (other than the sun) as seen at a particular location on earth. It is a few minutes shorter than the solar day, and unlikely to have biological relevance unless you think horoscopes have all the answers.

Strange cycles

Natural cycles like day, lunar month, and year are sufficiently fundamental that they have ready theoretical justification for inclusion in a model. The success of the model will determine the empirical support for that cycle in those data.

Bear in mind however, that if we arbitrarily decided to analyse based on, say, a 358-day period instead of 365, we'd often get fairly similar results, but even if it gave a better result 358 d would still be a strange cycle lacking theoretical justification. In most samples we could almost certainly find a better R^2 by tweaking a natural cycle's period; but unless that improvement were consistent over many samples we would end up with a series of cycles (e.g. 358, 373.3, 368, ... , etc.), describing in fact the samples and not the population. Citing them as real would be like over-fitting data and would imply a judgement deficit. Occasionally such strange cycles are claimed; but they appear to be inferred by methods not clearly laid out, and they lack proper statistical support for generalisation to a population. Such strange cycles are not even intriguing—they just present as error and poor judgement.

A cycle's period, to be credible, should have theoretical support, or clear empirical support, and preferably both. Cycles that differ only slightly from theoretically plausible conventional cycles need to be treated as (i.e. fitted to) conventional cycles unless authors can show statistically that they pertain to the population, not just to samples. The statistics to show that would generally involve a tendency or alternatively a trend amongst samples. Strange cycles cannot readily avail of the theoretical basis of established natural cycles (can't have it both ways).

The empirical identification of a period is not trivial. Here, like our predecessors Bliss (1958; 1970) and Batschelet (1981), we don't address finding the periods of variation. Instead, we assume periods *a priori*, relating periodic phenomena to well-established natural cycles. If you wish to identify a period empirically, you will probably use techniques like Fourier analysis or perhaps test a range with periodic regression.

6 PERIODIC REGRESSION: for cycles *(daily, seasonal, lunar, tidal, ...)*

Every biologist knows cycles exist. Therefore, we should all know about periodic regression, but few do. This book hopes to change that.

Periodic regression provides a way to analyse observations and describe cycles. Not only does it rigorously relate observations to cycles, but also (and therefore) provides a way to 'clean' data of the effects of certain circular variables where they are not the central interest but may be strong enough to interfere with analysis of another variable that is the central interest.

We are surrounded by periodic variation, e.g. daily or seasonal cycles that 'influence' temperature, rainfall, etc. We thus have circular variables (X) to which observations of (putatively) periodic phenomena (Y) can be related.

6.1   AN INITIAL EXAMPLE TO GET OUR BEARINGS

This example is simply to give a feel for the general process and what the results can look like. If you get more from it than that, wonderful, but remember there are important cautions (Sec. 6.4.6, 11.4) that are not explained in the example. (More examples are in the Appendices, Sec. 9.)

We'll use a simple short data set—sea surface temperature (SST, in degrees Celsius) from a small region west of Dominica, West Indies (Table 6-I). This data set is non-orthogonal (Sec. 6.4.6). We'll state what we want to know and express that in a regression model (Eq. 6-1). Then we'll convert (Table 6-II) the data to a form that allows periodic regression. We'll analyse it and interpret the regression output (Table 6-III), calculating where the peaks are, and what the amplitude of the function is. We'll plot (Figure 6-1) the regression results with the data.

We want to describe the seasonal (365d) cycle in SST, know where the maximum occurs, and what the amplitude of variation is. Time of year is DOY (day of year, 0-365; Sec. 5.3.1), and we include the second harmonic (Sec. 13.4) for a better fit.

The regression to describe what we want is of the form:

Eq. 6-1 $\text{Celsius} = B_0 + B_1 * \sin{}^` \text{DOY} + B_2 * \cos{}^` \text{DOY} +$
$B_3 * \sin{}^` (2\text{DOY}) + B_4 * \cos{}^` (2\text{DOY})$

where the grave (accent) mark (as in sin`DOY) indicates the proper transformations (Notation, Sec. 13.2). Other forms are possible: for example, we might have only one cycle (i.e. omit terms with B_3 and B_4), or we could add more harmonics, or a linear term to accommodate (test for) a supra-annual trend, or even categorical terms to recognise locations; but for now we'll keep it simple.

Table 6-I. Data: daily sea surface temperatures west of Dominica. (Data are from the box 61.48W to 62.01W by 15.07 N to 16.04 N, and were obtained from NASA Physical Oceanography Distributed Active Archive Center at the Jet Propulsion Laboratory, California Institute of Technology, http://podaac.jpl.nasa.gov/.) We show and analyse only a small excerpt (each 500th data point, resulting in a thinned data set of 17 cases), so that you can re-analyse the same data to follow along. This analysis ignores the year number, thus it assumes no year effect. Subsequent tables use these data to show the conversion steps.

X (the date)			Y
YYYY	MM	DD	Celsius
1989	8	6	28.65
1989	11	13	28.05
1989	12	29	27.15
1990	1	18	26.4
1990	2	5	25.65
1990	2	24	25.5
1990	4	1	26.55
1990	5	26	27.3
1990	7	29	28.05
1990	9	4	28.65
1990	11	8	29.7
1990	11	26	28.35
1990	12	20	27.45
1991	1	7	26.7
1991	2	22	26.25
1991	3	31	26.55
1991	5	21	26.4

We see that the regression model has terms that are sine and cosine of the date, but the data don't arrive in that format; date is in a common format (Table 6-I). Therefore, it needs conversion to trigonometric format (sine and cosine, Table 6-II) for the analysis (Table 6-III).

We begin by converting the dates to mathematically proper format (Sec. 3.4). We already identified our cycle of interest (year) so we convert dates to DOY (table, Sec. 16; formulas, Sec. 10). *Next*, to allow taking the sine and cosine transforms, DOY has to be further converted to angular format (radians). Thus, we converted date from common format to DOY to radians to (sin,cos); in other words, we converted X in common format to cyclic to angular to trigonometric format. This is done for each cycle, so if you are analysing just the main cycle (first harmonic) then you will need only one pair; we will include the second harmonic (Sec. 13.4), just to show how two simple curves add together to give a more complex one (Figure 6-1).

Thus, we get two pairs of transforms (more if we wanted more harmonics), all from the one single column (DOY).

Most computer programs expect angles to be in units Radians (2π radians equal one cycle), so we convert each DOY (as an angle in a $k=365$

unit cycle) to an angle in a $k=2\pi$ cycle (Sec. 3.5)by multiplying by $2*\pi/k$ = $2*3.141592654/365 = 0.017214206$, then take the sine and cosine. The second harmonic is half the cycle, 182.5 days, so for that we now go to DOY once more, multiply by 0.034428413, and take the sine and cosine.

Table 6-II. Sequence of transformations for an annual-cycle periodic regression. Dates as month and day common format (X_{common}) converted to Day-of-Year (DOY, 0-365; Table 16-I), thence to radians, then sine and cosine. Computerised sine and cosine functions typically expect input in radians. Rads=units$*2\pi/k$ for any cycle of length k units (Sec. 3.5.1); so, for the first harmonic calculate as cos or sin(DOY$*2\pi/365$), and second harmonics as cos or sin(($2\pi/182.5$)$*$DOY) or sin(($2\pi/365$)$*2*$DOY). See analysis in Table 6-III. The analysis uses the columns marked Y (given) and $X1$–4 (trigonometric representation of annual and semiannual cycles). In many situations only one harmonic would be included. The notation sin(2*Rads) is proper for 2nd harmonic (Sec. 13.4). Fitted values (not shown) form the regression line, plot-able against DOY.

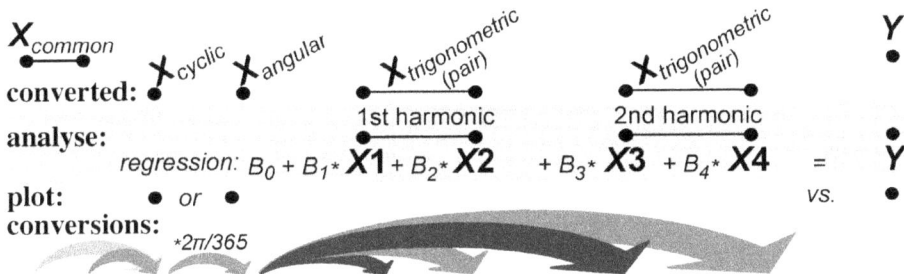

regression: $B_0 + B_1* X1 + B_2* X2 + B_3* X3 + B_4* X4 = Y$

			X1	X2	X3	X4	Y
		Rads	sin(Rads)	cos(Rads)	sin(2*Rads)	cos(2*Rads)	
MM	DD	DOY	R`DOY	sin`DOY	cos`DOY	sin`2DOY	cos`2DOY	Celsius
8	6	217	3.735	-0.560	-0.829	0.928	0.374	28.65
11	13	316	5.440	-0.747	0.665	-0.993	-0.116	28.05
12	29	362	6.232	-0.052	0.999	-0.103	0.995	27.15
1	18	18	0.310	0.305	0.952	0.581	0.814	26.40
2	5	35	0.602	0.567	0.824	0.934	0.358	25.65
2	24	54	0.930	0.801	0.598	0.959	-0.284	25.50
4	1	90	1.549	1.000	0.022	0.043	-0.999	26.55
5	26	145	2.496	0.602	-0.799	-0.961	0.276	27.30
7	29	209	3.598	-0.441	-0.898	0.791	0.612	28.05
9	4	246	4.235	-0.888	-0.460	0.817	-0.577	28.65
11	8	311	5.354	-0.801	0.598	-0.959	-0.284	29.70
11	26	329	5.663	-0.581	0.814	-0.946	0.325	28.35
12	20	353	6.077	-0.205	0.979	-0.401	0.916	27.45
1	7	7	0.120	0.120	0.993	0.239	0.971	26.70
2	22	52	0.895	0.780	0.625	0.976	-0.218	26.25
3	31	89	1.532	0.999	0.039	0.077	-0.997	26.55
5	21	140	2.410	0.668	-0.744	-0.994	0.107	26.40

The proper transformations here are: Rads=DOY$*2\pi/365$ (e.g. $217*2\pi/365=3.735$), then sin(Rads), cos(Rads), sin(2Rads), and cos(2Rads), which we label as sin`DOY etc. to retain identification. We now have proper sine and cosine transforms for each date, and for the main cycle (first harmonic) and the half year cycle (second harmonic), and we can now run the regression.

The regression is conducted as Celsius, the Y or dependent variable, $vs.$ the four proxy x-variables derived from DOY: sin`DOY and cos`DOY (1st harmonic), and sin`2DOY and cos`2DOY (2nd harmonic). (Notation: the grave mark ` indicates the proper sine, and second harmonic is indicated by a mathematically proper multiplication; see Sec. 13).

Table 6-III. Regression output, with interpretation, for data in Table 6-II. Regression function is readable from the coefficients as: CELSIUS=27.5-1.54*sin`DOY-0.307*cos`DOY-0.287*sin`2DOY-0.337*cos`2DOY. Phase (dates) and amplitudes are calculated from coefficients. Second harmonic (H2) has (by definition) 2 peaks in one main cycle, 0.5 cycles apart.

Regression Statistics

Multiple R	0.943
R Square	0.890
Adj'd R Sq.	0.853
Std. Error	0.447
Obs'ns	17

ANOVA

	DF	SS	MS	F	p[F]
Regression	4	19.417	4.854	24.294	1.12E-05
Residual	12	2.398	0.200		
Total	16	21.814			

	Coeffs	Std. Error	t Stat	p[t]	L.B. 95%	U.B. 95%
Intercept	27.54	0.12	233.68	0.00	27.28	27.80
sin`DOY	-1.54	0.18	-8.69	0.00	-1.93	-1.15
cos`DOY	-0.31	0.16	-1.91	0.08	-0.66	0.04
sin`2DOY	-0.29	0.14	-2.01	0.07	-0.60	0.02
cos`2DOY	-0.34	0.19	-1.77	0.10	-0.75	0.08

Interpretation of peaks (in units of x) and amplitudes (in units of y)

harmonics:	Phase (DOY)	Phase (mmdd.hhmm)	Amplitudes †
1st (H1)	262.32	0920.0738	1.57
2nd (H2)	111.77	0422.1833	0.44
2nd peak of H2	294.273	1022.0633	† contribution to Y

The typical overall null hypothesis for regression is that no coefficient differs significantly from 0. This regression is highly significant overall: $p[F_{4,12} \geq 24.94]=0.0000112$ in the null population, or under the null hypothesis (Sec. 11.1.7, 11.3.6). It has a high adjusted (conservative) R-squared of 0.85. Phases are interpreted separately for the two component cycles, which interact to create the overall phase (Figure 6-1).

Significance for each cycle is a subtle issue (Sec. 6.4.5, Sec. 10.2.1); the $p[t]$ values for their sine and cosine components are a poor guide. Residual SS in regressions (not shown) for 1st (DOY) and 2nd (2DOY) harmonics separately are 3.91 and 19.1, letting us evaluate each harmonic in the context of the full regression: DOY conditional on 2DOY has $Fimprv_{2,12}=41.8$ ($p=0.0000039$); 2DOY conditional on DOY has $Fimprv_{2,12}=3.78$ ($p=0.053$). Because the data are non-orthogonal (Sec. 6.4.6), Fcycle would only approximate these values.

The periodic regression function can be expressed as a curve on a flat plot (Figure 6-1). The first harmonic alone would be represented as a sine wave of period 365 d, and it has clear theoretical justification; inclusion of the second harmonic, having a period of 365/2 d, modulates the first harmonic. The second harmonic is often significant in real data, but that's a more utilitarian than theoretical justification. Beyond that we get to thinner ice: third, fourth (etc.) harmonics can be added, but usually with diminishing benefit and justification. We should not merely chuck in terms to improve a fit, because the ideal regression is one in which there is theoretical justification for each term.

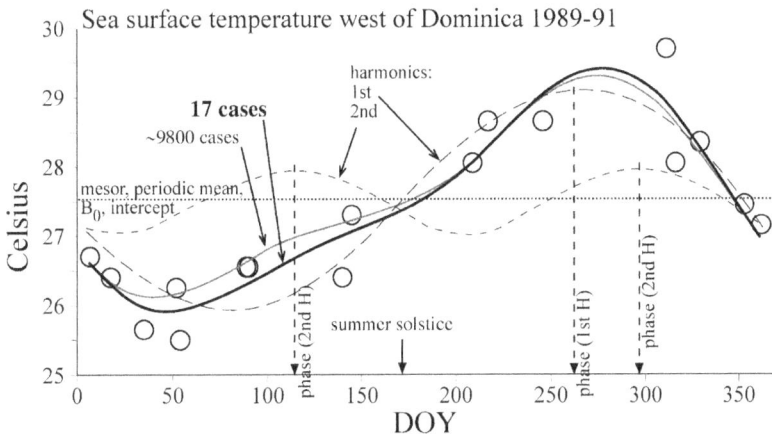

Figure 6-1. Data (circles), periodic regression and component harmonics. The regression (heavy curve) uses 17 cases, being every 500th case (circles) of the original set of ~9800 cases less missing values. The 17 cases give—remarkably—much the same result as the original ~9800 cases. The 1st and 2nd harmonic constituents of the heavy curve are shown as light dashed lines; both curves as presented include the mesor or intercept to keep them on the graph, but imagine all periodic components cycling around zero, then summed together with the mesor. The *mesor* is the value around which the whole periodic function oscillates; we obtain it from the intercept, but it has special meaning in periodic regression (later). Peaks due to first and second harmonics need not coincide with the overall peak. The second (or *n*th) harmonic peaks twice (or *n* times) per main cycle.

The peak in temperature is after, not at, the middle of the summer. Although some very bright people naturally assume that it should be at midsummer, they forget that matter stores heat; thus, temperature is due not only to that day's sunshine, but also to sunshine the previous day and (diminishingly) many days previous to that, all balanced against losses to the atmosphere and radiative losses to space. The inertial component retards the seasonal peak to somewhat after the seasonal peak in solar radiation. Here, the indicated peak temperature is in late September; counter-intuitively, it's much later than the summer solstice; but this is the sea surface and water stores a lot of heat, so it's reasonable that inertial effects would be great and move the peak later.

Likewise, with tides: the largest tide is not exactly on the full or new moon (when the gravitational forces of sun and moon are aligned and therefore have the strongest effect), but a couple of days later; that is sometimes referred to as the "tidal lag". All of that means that if you are ever looking for a relationship of some Y to day length, or to sunspot frequency, you shouldn't rely on simple correlation (which could fail simply due to Y being out of phase with X); instead you should model it as a cycle, using periodic regression to allow detection of relationships of any phase difference.

The graph (Figure 6-1) shows that the peak in the overall curve results from peaks in the first and second harmonics. These component functions can easily be separately calculated.

That's the general idea; it is not too difficult. You'll notice that periodic functions have the same value at the beginning and end of each cycle; and that each harmonic is a sinusoidal curve; the first harmonic has one peak and one trough in a cycle, and the second harmonic has two peaks and two troughs.

The rest of this section will, I hope, explain any steps that looked like sorcery.

6.2   INTRODUCING PERIODIC REGRESSION

Any Y that is potentially influenced by cycle(s) X (daily, seasonal, etc.) is a candidate for analysis by periodic regression. Some information on the biological examples referred to below can be found in papers via <http://www.ucs.mun.ca/~kbell/>. Worked-out examples are in Sec. 6.1 and Sec. 9.

Recall (Sec. 1.2, 4, 4.1, Figure 1-3) that cycles occupy two dimensions (at least). Our *angular* and *cyclic* representations (degrees 0-360, hours 0-24 of a day, days 0-365 of a year, etc.) don't directly support analysis, and they can blind us to periodicity.

6.2.1. MECHANICAL ANALOGY: THE KELVIN TIDE PREDICTER

A spectacular application is the analysis and prediction of tides. In the late 1800s, superior knowledge of tides conferred a tactical advantage, and the Kelvin machines, Lord Kelvin's tide calculators, (Figure 6-2) were the first success. Similar, room-sized, mechanical devices were standard equipment until not long ago: a main handle was cranked once to represent one day, and it turned a series of gear sets (to model the period and phase of each cycle), lever arms or plates (to model amplitude contributions of each cycle), then pulleys and a long cable throughout to sum the contributions of all cycles to tidal height.

A charming simulation by Bill Casselman at UBC shows how such machines (or any equivalent logic!) can sum multiple periodic constituents into a single curve. Tides have even been used to generate music, as on Casselman's site†.

(†www.math.sunysb.edu/~tony/whatsnew/column/tidesIII-0601/tidesIII3.html; Aug. 2008)

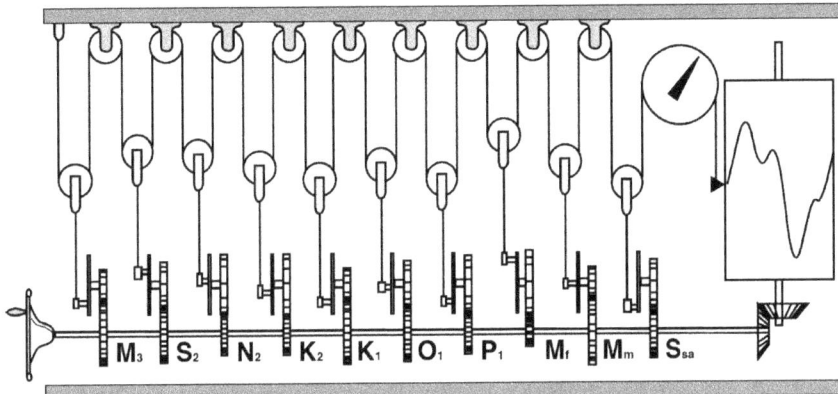

Figure 6-2. The Kelvin Machine for tidal calculation: a useful analogy for modelling by periodic regression. The main handle to the left represents one day per turn; attached gear sets each model a single mathematical constituent of the tide series. This model is a relatively simple one with only 10 periodic constituents (37 or more have been used since). For each constituent, representation of: [a] period is by its gear ratio with the main shaft; [b] amplitude (its contribution to the tide) is by the length of the crank that connects its spur gear to the pulley system; and [c] phase is set by angle of the spur gear. Tidal prediction has been done by electronic (as opposed to mechanical) computers since the 1960s. (Redrawn from various sources).

Periodic regression is analogous to Kelvin's machine. As in Kelvin's machine, the period of each cycle is treated as a given, and taken into account when determining sine and cosine from the original observations. As in Kelvin's machine, a single time index (the main shaft) can drive many different cycles.

6.2.2. OVERVIEW

We know cycles occupy at least two dimensions (you can't draw a circle in only one dimension). We know that most calculations require decomposing a cycle into its (paired) trigonometric functions, the sine and cosine, and we know how to do that (Sec. 3, 4). Now, therefore, we can attack a more ambitious and very useful task: quantifying the relationship of biological and other variables to daily, seasonal and other cycles. This kind of task is well handled by periodic regression.

Periodic regression relates linear Y scores to a circular or periodic X (e.g. DOY, sequential day-of-year, 0-365). Y is modeled as a sinusoidal cycle of pre-identified length or period (the length is not estimated statistically).

Some may ask "why not just use a polynomial?" A polynomial is not a periodic function because it does not return expected Y to the same value at the end of the cycle. The quick reader might now comment "but then how can the tides be periodic if they don't return to the same value at the end of the year?". In fact, the tides don't. What does return is each component, at the end not of each day or tide or year, but of its own cycle. Yes, the constituents not only can have different periods, indeed, they must: if two have the same period then they really are one constituent. The tide we observe is well explained as the sum of many sinusoidal components. Their sum doesn't return to the same point on an annual basis because the least common multiple of all the periods is many years.

Some more common approaches are no better. For example, dividing the year into seasons and doing an ANOVA like Celsius *vs.* season; the ANOVA cannot describe a pattern, partly because it has no way of "knowing" that the seasons are sequentially related. If trends are strong and your arbitrary divisions into seasonal groups aren't unluckily chosen, you can get a result indicating a significant departure from chance, but, as said, you will get neither an identification of the phase angle nor a model allowing prediction from the data.

Periodic regression has several applications useful in biology and medicine, for example:

[i]   describing periodic patterns, peaks, and troughs in data;

[ii] obtaining a robust mean value (a better estimate than the mean of all observations) where variation is periodic but observations are limited;

[iii] modeling circular and linear (or even categorical) independent variables together, even where the linear term is of principal interest but is obscured by cycles;

[iv] de-trending, or removing periodic constituents from, data to give residuals representing anomalies, or values de-correlated with their underlying cycles.

Thus, even in studies not primarily directed at cycles, periodic regression can help by correcting for variation attributable to those underlying cycles. Taking a very simple example, say you are interested in site-to-site differences like the temperature in two locations but you are not interested in any variation due to time of day, time of year, time of tide, i.e.: you want to know if there is a site-related difference. Periodic temporal sources of potential variation could affect the data, so you wouldn't want to attribute that to site. You have two alternatives: get a second team of people and equipment to measure the second location at the same time (expensive); or, sample both locations at non-identical times but reasonably representing the main cycles and then use periodic regression to de-trend the data. Thus if your data include variation due to time of day (or year, tide, etc.), it can be accounted for in a regression and this analysis can give you better resolution of the differences you are chiefly interested in. Let's say for example you have from each location a list of temperatures at various (haphazard sampling is often best because it is aperiodic) times of day and night. You can do a periodic regression of the temperature data, explaining it by the variables you are initially *not* interested in; effectively, the residuals are data *corrected* or de-trended for the cyclic trends (that you were initially *not* interested in), and you can now do an ANOVA or *t*-test on the regression residuals to compare the two locations. And as a bonus, not only do you have a far more reliable comparison of the two locations, but you now also have a model of the cycling of temperature, or of whatever *Y* was measured.

A periodic regression estimates one **mesor** for the entire regression, and then in each cycle estimates coefficients B_{sin} and B_{cos} from which (indirectly) we get **amplitude** and **peak** for each cycle. For more detail, see Sec. 6.4.4, but in brief these parts are:

i) **mesor** (M) or **periodic mean**: the value about which the modeled function cycles;

> ... like any intercept it is the value of *Y* *if* all *x*-variables, all the sines and cosines, were equal to zero—we say *if* because that in practise is impossible in periodic regression because there is no angle for which sine and cosine are both zero; so it is a special kind of intercept meaning the value about which *Y* cycles. We can think of the mesor as the 'periodic intercept' because of where we find it, or the periodic mean because of its utility.

ii) **amplitude** (*A*): the degree of variation about the mesor in the modeled cycle;

> ... calculated from sine and cosine coefficients by Pythagoras's theorem as $(B_{sin}^2 + B_{cos}^2)^{0.5}$) shows the maximum contribution (measured as departure from the mesor) to *Y* by the cycle.

iii) **phase angle** (also called **acrophase**, **peak location**, etc.): within a given cycle, the location where its contribution to *Y* is maximised;

> ... calculated from coefficients by arctan method, Sec. 4.3.2, or by plotting B_{cos} *vs.* B_{sin}. In any sinusoidal component, the minimum is 0.5 cycles away from the maximum.

These measures are substantially robust against the vagaries of sampling times, i.e. sampling at irregular times is well tolerated (in contrast to some applications of time-series analysis, and the Rayleigh test). Parametric significance of regression is valid where errors e_i are normally and independently distributed (Batschelet 1981). Presumably distributions other than the Normal could be substituted, so long as the regression significance is assessed using a distribution compatible with the distribution of errors.

6.3   CONCEPTUALISING PERIODIC REGRESSION

The key concept here is the sine and cosine pair. Analysis relies on decomposition of a circular *X* into sine and cosine, which become proxy *X*-variables.

Sine and cosine both have the same shape (sinusoidal), but are 90° out of phase with each other. They are products of simple geometric elements: if you cut a tube at an angle, then unroll the tube, the cut edge is a sinusoidal curve. More succinctly: a sine wave is the intersection of a plane and a cylinder. (From this you can be clever and correctly say that a straight line is a special case of a sine curve: one with amplitude = 0. This is a great way to slow down a good party.) The corollary is that you can roll up a sinusoidal curve and from the correct direction around the cylinder it will look flat, though tilted (unless amplitude=0).

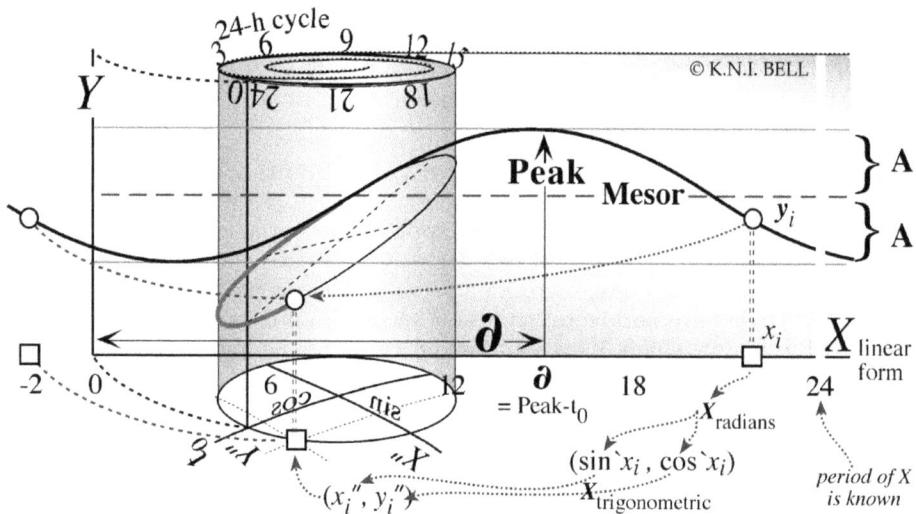

Figure 6-3. Conceptualising periodic regression: *Y vs.* circular *X*. A flat graph of periodic data in a wave-like pattern is "printed" onto a cylindric graph rolled across it, or vice versa. The *x*-axis of the flat graph is cyclic format of circular *X*, while the cylinder base has the $(\sin`x, \cos`x)$ coordinates of *X*, the transforms *that are essential for analysis*. The cylinder has radius 1.0 (unit circle). *Mesor, Phase angle* (∂) of the peak, and *Amplitude* (*A*) are marked.

Sinusoidal variation is therefore the simplest form of cyclic variation (because a plane is the simplest cut through a cylinder) and, like ordinary straight-line linear regression for non-cyclic variables, it is the reasonable starting point or default form in exploring and explaining periodic variation. Thus, once it is accepted that a variable is periodic, sinusoidality is the null expectation; no justification is needed for the sinusoidal form; only variations from that form need justification.

Figure 6-3 shows relates a flat *Y vs. X* system to a system where the circular *X* is properly expressed: *Y vs.* {*x''*,*y'*}, the cylindrical plot. (See *notation* (Sec. 13.3) and symbols (Glossary, Sec. 14).) Analysis will also use system like *Y vs.* (*x''*,*y''*) or *vs.* (sin`x,cos`x). One could of course use *x,y,z* space where *x* and *y* are the two linear (sin,cos) components of the cycle and *z* is the dependent variable, but we're accustomed to writing *Y* as the dependent variable so let's stick with that.

One can thus visualise periodic regression of one *Y* on one circular *X* as first rolling (Figure 6-3) a scatterplot of *Y* on *X* into a cylinder of circumference equal to the length of the cycle, e.g. one year, and then looking through the cylinder while rotating it (Figure 6-4) to the position where the data points can be seen from within the plane nearest which the data lie. If there is no cyclic trend in the data, they occupy a roughly horizontal band no matter from which direction they are viewed. So, just as the null hypothesis for a linear regression is "no trend" or "regression line slope not significantly different from zero", for a periodic regression the null hypothesis is "no trend" or "the plane, about which the data are scattered, is not significantly tilted in any direction".

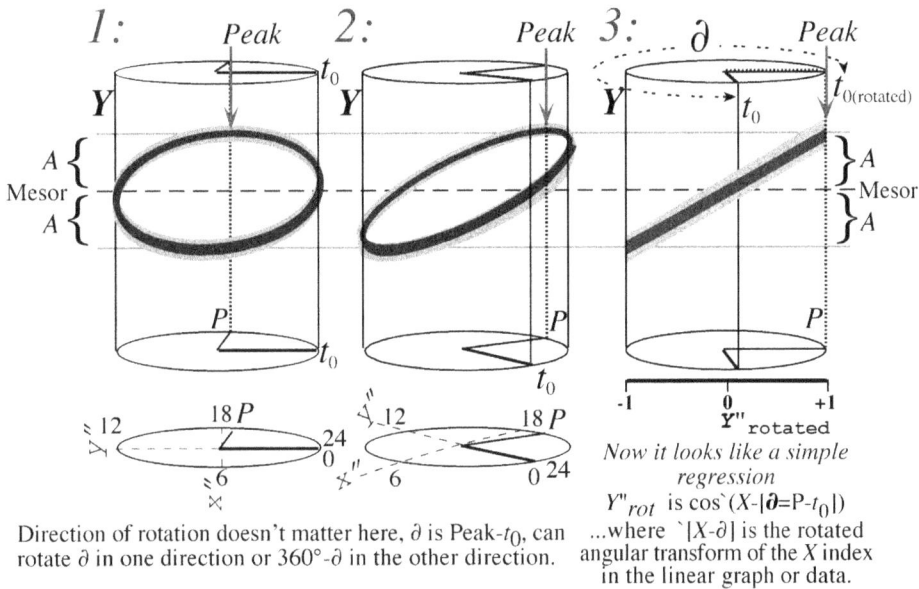

Direction of rotation doesn't matter here, ∂ is Peak-t_0, can rotate ∂ in one direction or $360°-\partial$ in the other direction.

Now it looks like a simple regression

Y''_{rot} is $\cos`(X-|\partial=P-t_0|)$

...where $`|X-\partial|$ is the rotated angular transform of the X index in the linear graph or data.

Figure 6-4. Rotating data to maximise correlation Y vs. $\cos`X$, or iterative fitting of $Y = M + B(\cos(R`X-\partial))$ Eq. 6-4. Data (Y) are the dark band around the cylinder. Figure 6-3 relates this to a flat plot of Y vs. X. $Y = f(x'',y'') = f(\sin`x,\cos`x)$, where X is circular, e.g. hour of day, represented 2-dimensionally as $(x'',y'') = (\sin`x,\cos`x)$. *1:* phase not identified; *2:* rotated to increase correlation of Y with Y''; *3:* rotated fully to peak, correlation maximised, phase identified and it looks like a simple regression. Y''_{rot} equates to $\cos`(X-\partial)$ where $`[X-\partial]$ is the rotated angular transform of the X index in the linear graph or data. **P** = acrophase, or phase angle of peak; **A** = amplitude; t_0 = arbitrary zero of cycle, e.g. 0000h of day. ∂ is the rotation† that maximises correlation of Y (dependent variable) with the Y'' axis ($\cos`(X-\partial)$). The cos component in rotated data is read from the $Y''_{rotated}$ scale [3]. Rotation also reduces the sine component to zero, tempting us to present periodic data as a simple linear regression; but beware the DF issue (Sec. 6.4.1). († It may seem counterintuitive that *peak* can be rotated $(X-\partial)$ to the *zero* of the cycle and still give a positive (not negative) slope on the cosine plot ('cylindrical') but it works because $\cos(0)=1$, and $\cos(180°)= -1$.)

To reiterate: Figure 6-3 and Figure 6-4 both illustrate that a sinusoidal curve is produced by slicing through a cylinder. The reverse also applies: if you wind a sinusoidal curve (with period = k) around a cylinder (circumference = k), you can then rotate the cylinder to show that the sinusoidal curve occupies a flat (but possibly tilted) plane in 3-dimensional (X'', Y'', Y) space.

With that basic concept, we can do the same thing algebraically. To make the data amenable to periodic analysis, the cycle (the circle around which the graph in Figure 6-3 is rolled) is decomposed into its two components: sine and cosine. Sine and cosine of the independent circular variable X are represented as circular coordinates x' and y'' (or y'' and x' if using the Polar angle mode instead of the Azimuthal) on which we plot the unit circle, with the unit circle being the base of the cylinder that we rotated to find the maximum positive correlation in visually fitting the lag.

In practice, periodic regression amounts to a multiple regression where the x-variables are sine and cosine (of angular transforms) of the original x-variable(s). It's an unusual transformation in regression for two reasons:

• *firstly*, we transform the x, not the y; and

• *secondly*, instead of getting one transform from one x, a pair (sin and cos) is generated; and they are used together. Effectively, we nominate a new set of axes, (sin,cos) of X, and call them (X'', Y''), for the index cycle (e.g. time of day).

The pair of transforms is required to convert the one angular measurement to a coordinate system, so consider them *an inseparable pair* for purposes of inference. I.e. it is absolutely *not* legitimate to drop the transform that does not come out significant, because dropping it will understate DF for the regression and will therefore cause F to be greatly overestimated (Sec. 6.4.1).

Periodic regression is little (if at all) affected by asymmetry of sampling effort, although complexities such as departures from sinusoidality may go undetected if there are large sections of the cycle that are without data.

The mesor can be described as the value around which the function is distributed, or the center of the ellipse formed by the intersection of a plane and a cylinder. The subtle but valuable aspect of the mesor is that it represents an *ideal* mean value of Y for an integer number of periods of the function. In contrast, the actual mean of the data would only equal the mesor if all points were equally sampled and equally spaced with respect to the cycle (i.e. if the *mean vector* of sample density had zero length). The mesor thus has a very useful quality: under the assumptions that the pattern of variation is sinusoidal and that the correct period length was applied, the mesor *provides a better estimate of the central tendency of the parent population than does the mean of all measurements*.

6.4 APPLYING & INTERPRETING PERIODIC REGRESSION

You'll have seen that a periodic regression is like a wave having both phase and amplitude. We know beforehand, or choose, what period(s) we are analysing for, whether a day or a year, etc. There are two approaches to fitting a periodic regression: the standard form (also called "Fourier form") and the iteratively-fitted "cosine" form. Both must estimate two quantities for each cycle, but the phase lag is estimated using trial-and-error in the "cosine" form while the standard form does it explicitly.

These techniques are explained Bliss (1958; 1970), Batschelet (1981) and earlier versions of this book. ANOVA is a legitimate technique with periodic data, provided residuals are independently and normally distributed (Batschelet 1981). Least-squares is a standard method for fitting regressions. Examples are given in some of the author's publications (Bell 1994; Bell et al 1995; Bell 1997; Bell et al 2001; Cowley et al 2001; Bell 2004; Bell 2007), above in Sec. 6.1, and in Sec. 9.

I will draw on an example from my study of tropical anadromous gobies—these are neat little fishes that at one stage of their life cycle migrate from the sea into (recruit to) rivers—where the age-at-recruitment (AAR) was the dependent variable (Y) of interest. Following Eq. 6-3, we'll use as example the regression (Bell et al 1995) for AAR of S. *punctatum* vs. day of year, with significances below coefficients, is:

Eq. 6-2: $AAR = 83.835 \quad + 10.1\sin\grave{}DOY \quad - 0.213\cos\grave{}DOY;$

$\qquad\qquad B_0 \qquad B_1 \; (p[t]=0.0001) \quad B_2 \; (p[t]=0.924)$

$\qquad\qquad R^2=0.28, \; n=117, \; p=0.0001$

where DOY is day of year and the grave mark ($\grave{}$) denotes (see Notation, Sec. 13) the proper conversion to radians etc. The $p[t]$ values operate as a pair: neither alone is reliable for the significance of a cycle (Sec. 6.4.5).

Periodic regression involves appropriate transformations of X, using them in regression, and interpreting the output. The transforms are key; they're easy enough but trivial errors are also easy, so make a habit of checking transforms with series (like 0, 30, 60, ... , 360 degrees) that you can plot to verify the proper sinusoidal shape and correct period.

6.4.1. STANDARD FORM, EXPLICIT (SIN`$X$, COS`X)

The standard form, also called the "Fourier" form (Bliss 1958), is mathematically equivalent to the "cosine" form (Sec. 6.4.2). The standard form is more efficient than the iterative 'cosine' form, because it finds the phase lag ∂ explicitly rather than iteratively; that also makes it less likely to get you into DF difficulties (Sec. 6.4.6). The standard form uses the proxy X-variables sin`$X$ and cos`X (the grave symbol ` signifies proper transform, see Sec. 13.1) that are trigonometric transforms of the circular X variable.

A simple periodic regression looks like:

Eq. 6-3: $Y = B_0 + B_1 \sin`X + B_2 \cos`X + \varepsilon_i$

where B_1 and B_2 are coefficients for the paired trigonometric transforms sin and cos of the circular X, the grave symbol ` indicates a proper transform of X prior to taking sine etc., and B_0 is the intercept that in the periodic regression context has special meaning recognised by the names *mesor* or 'periodic mean'. A normally-distributed error ε, or residual, is generally present even if not noted. Eq. 6-3 can be solved directly using $\sin`X$ and $\cos`X$ as proxy X variables (once their values are calculated, the regression is like a multiple regression). Both the phase angle (lag) and the amplitude can be easily calculated from the regression parameters.

Surprisingly few data may suffice for periodic regression; it really depends on the variances within the data. Batschelet (1981, p. 164) says "... to obtain meaningful results ... a minimum of $n = 6$ equally spaced [data points] will suffice... however, if the intervals are not equal, at least $n = 8$ data points are required." He also points out that reliability of estimation of the parameters will suffer if all the points are clustered in a single region of the cycle, i.e. severely non-orthogonal, Sec. 6.4.6).

In principle however, if error is zero then only three points should be needed to fit a single cycle, because in the cylindrical model the regression is the intersection of a cylinder and a plane; then remember from basic geometry that two points can specify any straight line, three can specify any flat plane. Thus, the presence of error in data is the reason for requiring more points than three to specify a periodic regression. Batschelet's recommended number of points was a practical minimum: in some cases errors (variance) will be such that even "8 data points" will be insufficient. (A periodic regression with many data should show little change in peak and amplitude when re-run on randomly chosen and reasonably large subsets; e.g. Figure 6-1).

Frequently we will have data on only one half of a cycle, as for example a 24h cycle for which we only have collected only in daylight. If those data are enough to define a cycle, the next question is whether that cycle could or should be presumed to continue in the un-sampled portion of the cycle. Some might argue that day and night should not fit the same function. Therefore, if you were prepared to assume (as opposed to wanting to show) sinusoidality throughout the cycle, then you could declare that assumption and use the daylight data to model the entire cycle; whether you were willing to use that function to predict night-time values, or to restrict your predictions to the daytime period that the data came from, would be your judgement call. In fact, as the sinusoidal curve is the simplest possible curve (being the intersection of a flat plane and a cylinder), it is the parsimonious assumption or null hypothesis for any cyclic variation. Anybody objecting to that would find themselves arguing against Occam's Razor.

6.4.2. "Cosine" form: iterative fitting by rotation of X

You will rarely need to use the cosine regression form, also called "cosinor analysis". If you're not sure you want this form, you don't. It can be shown to be mathematically equivalent to the standard form (Eq. 6-3), but it's cumbersome and if you're not careful you'll exaggerate significance (Sec. 6.4.6). It is however more intuitive and may be useful for teaching purposes. If you really want this form, the best way to get it is to solve it in the standard form, Sec. 6.4.1, then convert to "cosine" form as described in Sec. 6.4.3.

The iterative method is fine (though tedious) for estimating a predictive equation, but it carries a serious danger: it is too easy to forget to correct the DF and therefore drastically overestimate F and significance (see cautions, Sec. 6.4.6). As well, simultaneous iterative fitting of multiple cycles quickly becomes impossible.

Sinusoidal variation (with frequency = 1) in a Y variable over a cycle can be expressed as

Eq. 6-4: $\qquad Y = M + A(\cos`(X–∂)) + ε$

where $R`X$ is (see Notation, Sec. 13) an angular transformation of an X (independent) circular variable (DOY, time of day, etc.—for which the period is known beforehand). **M** is a mesor (or periodic mean), **A** is the amplitude, and $∂$ is the phase angle or the distance between the peak and the arbitrary zero of the X cycle.

To solve the regression iteratively, first guess a number of lags ($∂$) to try, and subtract them from the $`X$ (the $X$ in an angular format like radians), and take cosines taken of each lag-transformed $`X$ (yielding columns $\cos\{`X–0.5\text{rads}\}$, $\cos\{`X–1.0\text{rads}\}$, $\cos\{`X–1.5\text{rads}\}$, ... etc.). Secondly, plot Y against each lagged set of $\cos\{`X–∂_{guessed}\}$ data and obtain r (Pearson product-moment correlation coefficient). Make a new list of lags $∂_{guessed}$ with their corresponding r. Find the appropriate lag, that giving the best positive correlation of Y with $\cos\{`X–∂\}$, by plotting **r** vs. $∂$; don't use **r²** because that will not help you identify the *positive* correlation. Fitting the lag amounts to choosing the one that reduces the $\sin`X$ component to zero while maximising the $\cos`X$ component. Usually, $r$ varies smoothly so the best $∂$ can be judged at the maximum of the curve of points, and if additional $∂$s need to be evaluated it is obvious. Least-squares regression is then used to describe $Y$ (size-, age-at-recruitment, etc.) in terms of $\cos(R`X–∂)$. Ensure that, if estimating significance, you have 2 penalty (numerator) DF per cycle. Check by plotting fitted and observed values vs. X (e.g. time, angle, etc.).

6.4.3. Converting between standard and "cosine" form

For completeness, we show that a regression solved in the standard form (Eq. 6-3) can easily be rendered into 'cosine regression' form Eq. 6-4 or Eq. 6-5,

Eq. 6-5: $y_i = M + A(\cos`(t_i-\partial))$,

by substituting M, A and ∂ as in Table 6-IV. Mesor or periodic mean M = B_0 from the standard form regression obtained above; A = amplitude by Pythagoras' theorem from the coefficients for sine and cosine; ∂ = phase angle by the arctan method from the ratio of sine and cosine coefficients and a quadrant correction (see Sec. 4.3.2). ∂ can be expressed as radians or degrees or a fraction of the cycle, or in the same units as X (e.g. DOY, k = 365). A regression in the cosine form can also be converted back to standard form Table 6-IV.

Table 6-IV. Converting between standard and cosine forms of periodic regression (using the example Eq. 6-2). Phase angle and amplitude are estimated from sine and cosine coefficients (Sec. 6.4.4, 4.3). Underscored values are converted from previous block.

Given (Standard)	AAR = 83.835 + 10.1sin`DOY - 0.213cos`DOY		
Standard	mesor	Bsin	Bcos
copy:	83.835	10.1	-0.213
Cosine from Standard	mesor	A	Phase angle (∂, of 365)
	83.835	10.102	91.2 deg = DOY 92.47
	Y= 83.835 + 10.1 * cos`(X_{DOY}−92.471)		
Standard from Cosine	mesor	Bsin=A*sin`∂	Bcos=A*cos`∂
	83.835	10.1	-0.213
	Y= 83.835 + 10.1 * sin`X +-0.213 * cos`X		

The cosine regression form leads naturally to a 'cylindrical' plot of Y vs. cosine of rotated and transformed X (which can be labelled $Y'' = \cos`[X-\partial]$, see Sec. 13). It looks like an ordinary straight-line regression (Figure 6-10), so it can be useful for convincing reluctant people that there really is such a thing as periodic regression.

6.4.4. INTERPRETING REGRESSION OUTPUT: MESOR (OR PERIODIC MEAN), PHASE ANGLE, AND AMPLITUDE

Three key parameters (introduced briefly in Sec. 6.2 and Figure 6-3) of a periodic function are: the **mesor**, M, the value about which a periodic function cycles; the **amplitude**, A, giving the maximum positive and negative contribution of the cycle to Y; the **phase angle** (or acrophase angle), ∂. Phase angle is the angle at which the contribution of the cycle X to Y is maximised. In a function with multiple cycles the overall Y_{max} results from all, and need not coincide with the phase angle of any.

The following descriptions of phase angles and amplitudes all relate to each cycle and its contribution to Y.

Mesor or periodic mean M (in units of Y)

The **mesor** or **periodic mean** is the value about which the function cycles. Of the two synonyms, the word "mesor" (originally an acronym for Midline Estimating Statistic Of Rhythm) is a little less intuitive than "periodic mean", but is less cumbersome if you need to write it many times.

The mesor is read directly from B_0, or the intercept (although it is a special kind of intercept), in a strictly periodic regression. That applies in a periodic regression that consists of cycles only, because the addition of a non-circular X variable will affect the intercept and the intercept in that case should not be called a mesor, but an operational mesor could be readily calculated for any set of held-constant values of the non-circular Xs.

The mesor is *not* an average of the data. It is an estimate of the central tendency of the parent population Y, assuming it to be sinusoidally dependent on X, and assuming Y to have been randomly sampled at each sampling time, even though the sample times need not have been evenly distributed over the cycle. If data are fairly evenly spaced with no missing values, then the average will be close, but the mesor is a far superior measure of central tendency than the average. You can think of the mesor as an average corrected for the cycle and without bias due to sampling times; or think of it as an average estimated for the process sampled rather than of the data gathered. The robustness of the mesor makes periodic regression useful for otherwise-difficult comparisons, such as where a Y (like temperature) is affected by time of day yet where it is impractical to measure Y at exactly the same time in two locations. The mesor or periodic mean allows such comparisons, even where data are not collected at the same times in both locations.

Table 6-V. Superiority of the periodic mean (mesor). Uneven representation greatly affects the average (arithmetic mean), but the mesor is robust. The *average* approximates the theoretical mean value (0) of a sine when sampling is even, but departs when sampling is uneven. The *mesor* is not so affected by unevenness. Data: 13 points, 0°-360°, evenly spaced in a cycle compared to 18 points with increased (uneven) representation in part of the cycle.

Situation	arith. average	mesor	remarks
even representation (original)	-3.592E-17 (≈0)	0	av. ≈ 0 = mesor
uneven (with repeats)	-0.096225	0	av. affected
Effect of unevenness:	-0.096225	0	mesor superior

Although increasing sampling over a portion of a cycle or range X will bias an average of X, in a regression it only increases accuracy at some locations and therefore improves overall accuracy. This is easy to demonstrate: construct a sinusoidal series with evenly spaced points; do the periodic regression (you won't get significance estimates because there is no error in Y and therefore $SS_{residual}=0$, but you'll get the function); determine the mesor and arithmetic mean, and note that they are very close. Then, repeat some of the points in one part of the cycle, and again calculate the mesor and arithmetic mean, and see (e.g. Table 6-V) that the average has shifted but the mesor is unchanged. In determining the mean value of Y over cycles, the mesor is robust despite asymmetry, whereas the mean is vulnerable.

Essence of the mesor: the mesor or periodic mean predicts what the mean would be if samples were evenly spaced over the entire cycle. For instance if samples incompletely or unevenly cover all times of day, and are needed as an index of a quantity such as plankton, and the plankton counts vary with time of day (or both time of day and time of year), then the mesor is useful as a predicted mean to compare plankton abundances among different locations. The mesor reliably has these properties only in regressions that are purely periodic (i.e. without linear or category terms).

Amplitude (in units of Y)

Here, *differing* from Bliss's (1958) terminology (Bliss calls it the *semi-amplitude*) but *consistently* with that in physics and electronics, amplitude A is the maximum positive or negative departure of predicted Y from the mesor. The maximum (Y_{max}) of the function is therefore $B_0 + A$, the minimum (Y_{min}) is $B_0 - A$, and the entire **range**, $Y_{max}-Y_{min}$, equals 2A. It helps to think of the situation in terms of right triangles, with the coefficients B_1 and B_2 (for sin and cos) being the lengths of two sides; A is then the hypotenuse (the line joining the origin with the coordinates (B_1,B_2) and can be calculated (Batschelet 1981, from his eq. 8.2.3) using Pythagoras' theorem:

Eq. 6-6: $A = (B_1{}^2+B_2{}^2)^{0.5}$

or, for the current example of age-at-recruitment (Bell et al 1995),

Eq. 6-7: $A = (10.1^2+(-0.213)^2)^{0.5} = 10.102$ units of AAR (days) .

The amplitude is the maximum positive or negative departure from the mesor; the predicted range is from the mesor–A to mesor+A, therefore 2A. (Some call A the 'semi-amplitude' and use 'amplitude' to mean what we write as 2A; see under Sec. 6.4.4.).

Amplitude indicates relative importance of a cycle

Often a regression will contain more than one cycle, and it will be useful to know which cycle is the most important.

Each cycle in a periodic regression has two significance values: one each for the sine and cosine components. There is a need to be able to integrate or amalgamate the information associated with the sin and cos components to show something about the importance of one cycle among several in a regression.

The amplitude is by definition in units of Y. This means that the amplitudes of each component cycle indicate the effect of that cycle on Y, and can be used to compare the relative importance of different cycles within a regression.

Phase angle (in units of X)

The phase angle (or acrophase angle, peak angle, angular location of the peak P) is the value of the cycle X where its contribution to Y is maximised. Phase can be seen in plots, approximated to quadrant (Figure 4-5) from the signs of (B_{sin}, B_{cos}) coefficients in a regression like Eq. 6-3, or calculated by algebra provided by Batschelet (1981). Phase angle is ∂ (delta) in units of X after the nominal zero of the cycle (x_0 or t_0):

Eq. 6-8:   Phase angle = $x_0 + \partial$; so where $x_0 = 0$, then $P = \partial$.

(That may seem a somewhat stupid thing to write, being little more than an identity, but it does remind that we can 'rotate' the Y data in the cyclic format X (Sec. 3.4) by adding a constant to all x_i.)

We first estimate an uncorrected phase angle as ∂' units after x_0 as:

Eq. 6-9: $\partial' = \arctan(B_1/B_2)$

where B_1 and B_2 are the regression coefficients (as in Eq. 6-3). The uncorrected phase angle ∂' is corrected as:

Eq. 6-10: $\partial = \partial' + QC$,

QC being a *quadrant correction* (Sec. 4.3). ∂, at this stage of calculation typically in radians, is easily converted (Sec. 3.5) to °, DOY, etc.

Example: obtaining mesor, phase angle, and amplitude from regression output

Our example is Eq. 6-2. We immediately see that the sine coefficient is large and significant, compared to the cosine coefficient. What does that mean? ... simply that the peak is near one of the axes; it is *not* a reason to drop one coefficient and re-run the regression—that would easily hide a penalty DF, leading to overestimation of F and significance (Sec. 6.4.2, 6.4.6).

Mesor or **periodic mean** is read directly from B_0: here, it is 83.835.

Amplitude, A, and range: following the Pythagorean formula $A = (B_1^2+B_2^2)^{0.5}$, $A = (10.1^2+(-0.213)^2)^{0.5} = 10.102$ units of AAR (days). The predicted range of AAR in the example is $2A = 2(10.102$ days$) = 20.204$ days.

Phase angle or **acrophase** can be roughly located by inspecting the signs of the (sin,cos) coefficients for the cycle, or calculated explicitly.

Quick inspection: the phase angle is clearly in the second quadrant, between 1/4 and 1/2 way through the cycle; how? ... by signs (+,-) B_1 and B_2 (coefficients of sin and cos components of the cycle) showing the quadrant (Figure 3-2, Figure 3-3). Then, because B_{sin} is much (ca. 50 times) larger than B_{cos}, the peak is much further along the X axis than the Y, so it is just over 1/4 way through the cycle.

Explicit calculation: the phase angle is ∂ units after the nominal zero of the cycle. $\partial = \arctan(B_1/B_2) + QC = \arctan(10.1/-0.213) + 0.5$cycle = 91.2 degrees, equivalent to DOY92.475, April 02. (To choose the angular units of the lag, just set your calculator to "degrees" or "rads" mode before taking $\tan^{-1}$).

Alternatively, in radians, arctan(-47.418)= −1.55, add QC of π (because 0.5 cycle=π) to get true ∂ = 1.59 rads; multiply by $365/2\pi$ to convert from radians to DOY, yielding the (same) phase angle of 92.475 days. Thus, you can do more or less of the calculation in the units you prefer. Although we transformed to sines and cosines using radians, they would have been the same whatever angular measurement unit was used for calculation—just make sure the calculator is set to the same units.

The cosine form Eq. 6-5 uses the phase angle as a term, and can express the phase in different but equivalent ways, e.g., where t is the time index (in rads, or degrees, or DOY) of the observation:

Y = 83.835 + 10.102cos($[t_{RADS}-1.59rads]$), using radians, or

Y = 83.835 + 10.102cos($[t_{degrees}-91.2°]$), using degrees, or

Y = 83.835 + 10.102cos($[t_{DOY}-92.475$ days$]*360/365$), preserving

   DOY units with conversion to degrees.

6.4.5. SIGNIFICANCE OF A CYCLE IN REGRESSION WITH OTHER VARIABLES

Overall significance of a regression is given by the $p[F]$ value. In a regression with only one cycle and an intercept, the cycle's significance is the same as the overall significance. In regressions with multiple cycles or other variables however, significance of a cycle is a subtle issue; $p[t]$ values of sine and cosine components are unreliable, with the lowest $p[t]$ easily 0.1 to 10 times the true probability of the cycle. We therefore need a way to obtain the significance of one cycle or harmonic in the context of other regression terms; the joint probability of its sine and cosine components given the other terms in the model.

To estimate in-context significance of *m_added* variables (e.g. C & D), a classic method (W.G. Warren, pers. comm.) requires two regressions: [1] a "full" model (e.g. with terms A,B,C & D) and [2] a "short" model lacking C & D (i.e. with A & B). The two models' residual SS and residual DF are used to calculate an F which I'll call *Fimprv* (available as a macro in Sec. 10.2.1, and applied in Sec. 6.1). *Fimprv*, on DF for *m_added* (ΔDF = 2 for a single cycle) and the residual DF of the full model, is:

Eq. 6-11: $Fimprv[m\_added, DF_{full}] = (\Delta SS/\Delta DF)/(SS_{full}/DF_{full})$

         $= ((SS_{short} - SS_{full}) / DF_{short-full}) / (SS_{full}/DF_{full})$

Alternatively, because $t^2 \approx F$, I suggest *Fcycle*, an average F. It matches *Fimprv* when the analysis is orthogonal (Sec. 6.4.6); otherwise it is an approximation, but safer than the lowest $p[t]$. If you (e.g. as a reviewer) have a regression table but not the data, *Fcycle* can be calculated from the t values for the cycle's sine and cosine regression coefficients, and its DF are 2 (for the cycle) and the residual DF of the regression:

Eq. 6-12: $Fcycle_{2,DFresid} = (t_{sin}^2 + t_{cos}^2)/2$

6.4.6. CAUTIONS

Non-orthogonality

Orthogonal designs generally increase efficiency in analyses, and give exactly (vs. approximately) least squares parameters. Structurally, orthogonality means the products of the elements of the X variables, here $\sin^{\cdot}X_i$ and $\cos^{\cdot}X_i$, sum to 0, as when data are evenly spaced over a cycle. That however is rare in real data, and more difficult with multiple cycles. A series of 12 monthly data is orthogonal only if pretending months are equal in length, and non-orthogonal if correcting for that. 365 d has only one pair of integer factors: 5 and 73—all others yield fractional sample intervals, which are inconvenient or unworkable. Orthogonality is thus either expensive or an illusion when analysing over the year.

Recall Batschelet's (1981) recommendation of $n{\geq}6$ data *equally spaced* or $n{\geq}8$ data *reasonably dense* over the cycle X. How serious a problem is non-orthogonality? We can explore this with synthetic (simulated) data. A periodic function $Y{=}B0{+}B1{*}\sin^{\cdot}X{+}B2{*}\cos^{\cdot}X$ was randomly sampled at values (times) of X chosen by uniform random numbers scaled to k, the length of the cycle. That gave exact values of Y, so random error (approximately standard normal) was added. 100 sets of 36 values of $\sin^{\cdot}X$, $\cos^{\cdot}X$ and Y were generated. On each set, regressions based on $n{=}36$, 12, 8 (Batschelet's advice), and 5 (rule of thumb $n{\geq}2m{+}1$) were run. Coefficients were plotted against non-orthogonality (measured as $\sum\{\sin^{\cdot}X_i\cos^{\cdot}X_i\}$). This was all repeated for Y errors scaled up by a factor of 8. Surprisingly, results indicated no loss of precision due to non-orthogonality from randomly-timed sampling, so it is not shown to pose a practical problem. Incidentally, estimates seem reasonable with n above the general rule of thumb $n{\geq}2m{+}1$.

Accepting non-orthogonal random or haphazard timing of sampling eases logistical constraints. It improves the chance of distributing data reasonably over the smallest cycle that could have an effect, and that is important for generalisation (see confounding, Sec. 11.4) even if that cycle is neither of direct interest nor analysed. Temporal (or index) resolution should increase in some relation to n. Least-squares estimates can be verified if needed by nonlinear fitting. Random subsetting of data can help probe robustness of conclusions.

Appropriate DF: each cycle uses 2 DF

Degrees of freedom (DF) are not used in fitting a function, but they are crucial in estimating its statistical probability. To refresh on SS, MS, F, DF, etc., see Sec. 11.1.7, 11.2.2, 11.3.7.

You can think of DF_{total} as the total credit available for estimating parameters and making a statistical conclusion. $DF_{regression}$ is then the credit used up in estimating the function, and $DF_{residual}$, which is $DF_{total}{-}DF_{regression}$, is the credit remaining to support significance of the F value you calculate from your data. Errors in DF therefore are important because they can lead to erroneous conclusions.

The appropriate regression DF is 2 per cycle (Table 6-VI), but your statistics program can't see DF you didn't show: if you conduct a cosine form (Sec. 6.4.2) regression after pre-fitting or assuming a phase, that pre-fitting costs 1 DF, but the regression software can't see the bill. 'Seeing' only a form like "y = b + mx" it assigns only 1 regression DF, leaving you with a problem to fix: $MS_{regression}$, $MS_{residual}$, and F are all wrong (F can be double); both MS must be re-calculated, using correct DF, from $SS_{regression}$ and $SS_{residual}$, to obtain a correct F and $p[F]$.

The DF error is avoided by the standard form of periodic regression (Eq. 6-3), because it conserves the number (and dimensionality) of X variables through the analysis, as illustrated in Table 6-VI.

Table 6-VI. Appropriate DF when using either standard (explicit) form, or cosine (iteratively fitted) form. Example is the norepinephrine data in Sec. 9.2. If you lag X by fitted or guessed ∂ and then present "Y vs. $B_0 + B_1 cos`(X-\partial)$" to the stats software it will have no way to "see" that you fitted ∂ beforehand, and it will not assign the appropriate DF; while the regression may be correctly estimated, its significance will be inflated.

	standard form (Eq. 6-3) $Y=B_0+B_1 sin`X+B_2 cos`X$		"Cosine reg." form (Eq. 6-4) $Y=B_0+A*cos`(X-\partial)$	
n:	12		12	
$DF_{regression}$ as seen by software	2	because there are 2 X variables: sin & cos	1	because only 1 X variable, cos`(X), is shown to the stats routine
$DF_{residual}$ as seen by software	9	because $DF_{residual}$ = n-1-$DF_{regression}$	10	because $DF_{regression}$ is incorrect
lacking	0	lag not pre-fitted, DF already correct	1	for having pre-fitted the lag
What the DF should be for both forms:				
$DF_{regression}$	2	= number of variables estimated beyond intercept		
$DF_{residual}$	9	= n - $DF_{regression}$ - 1 , or n-m-1		

Caution 1 is therefore: to avoid DF errors, use the standard form (Eq. 6-3) rather than the cosine form (Eq. 6-4).

Don't split sin`x from cos`x

Caution 2: If you want accurate estimates of significance, *never* discard sin(x) or cos(x). *Even if the coefficient for one of them is zero*, that only means that the phase angle is close to 0°, 90°, 180° or 360°, but you couldn't have known that without using both sin and cos in the analysis, so you still owe 1 DF. This mistake (deleting the less significant component) is tempting and common, but such "stepwise" logic will cause the $DF_{regression}$ to be underestimated, $DF_{residual}$ to be overestimated, and thus F overestimated, just as with the easy error in the cosine form (Sec. 6.4.2). Caution 2.1 is therefore: never use automated stepwise regression for periodic regression.

Intercorrelated *X*s (e.g. de-trending temperature)

<u>Caution 3</u>: when using two or more X variables in any regression, independent variables should not be strongly intercorrelated. There is no such thing however in general experience as a correlation of zero, so "how strongly is too strongly" becomes a judgement call; but perhaps $r^2 > 0.15$ might be cause for concern.

Beware using a periodic variable (one that responds to a circular independent variable) as independent variable along with circular variables they already respond to. For instance, if interested in both temperature (well known to be a periodic variable) and day of year (DOY) in relation to some Y, ensure that temperature and time of year are not strongly correlated (but they usually are). If they are correlated, you cannot use them together in the same regression, unless you de-trend one with respect to the other. Without de-trending, you risk mis-attribution of variation, and you may have a collinearity problem. I.e. your results may be less useful than they look.

The solution to intercorrelated variables is to either drop one, or de-trend one with respect to the other—you can't de-trend the time of year (congratulations if you can answer why[e]) but you can de-trend the temperature with respect to the date. The practical way to do this is to do a regression of temperature *vs.* time of year (Celsius = B_0 + B_1sin˙DOY + B_2cos˙DOY) and obtain the residuals, which will be the portion of temperature not predicted by the regression, or in other words the temperature anomalies.

We can then use the de-trended temperature, or temperature anomalies, in a regression. Let's suppose our *Y* is something like stroke recovery rate, or pollination rate, etc. in a regression $Y = B_0 + B_1$sin˙DOY $+ B_2$cos˙DOY $+ B_3$*Celsius_Anomalies). Comparison of the significances of the temperature anomalies and of the other predictors will indicate which is the stronger contributor to *Y*—either temperature itself or something else that depends on the time of year. You may then compare a model using temperature alone with a model using time of year, or a combined model. It will be up to you to judge whether either or both variables are important enough to be included in the regression model you publish.

6.5   PLOTTING AND PRESENTATION

Plotting is a key communication method for quantitative data. Plotting of periodic regression lines however is not conveniently handled by any software I know of at present. Nevertheless, it's easy to improvise.

---

e Why: *firstly*, date is not a dependent variable, so it makes no sense to try to de-trend it; *secondly*, the date is a circular variable which in analysis must be two components and you can't do a regression with 2 *Y*s, so you can't de-trend the date.

Most stats programs will draw the line for a straight or polynomial regression, but cannot draw a periodic regression. So here's a simple basis for several kinds of plots: create a column of fitted Y values ($\hat{y}_i$, the expected value of Y at an observed X). If there are many cases, fitted values will result in many redundant points, so you might prefer to use a short series of predicted values (expected value of Y at arbitrary Xs), e.g. 10 values spread evenly over the cycle to enable tracing a smooth curve. Many programs offer fitted and predicted values, but it's just as simple to calculate them from the regression equation. Plot the fitted and/or predicted values against the X variable to present the regression line, and plot the observed values with it. To make a cleaner plot, replace the many fitted values with a drawn-in line.

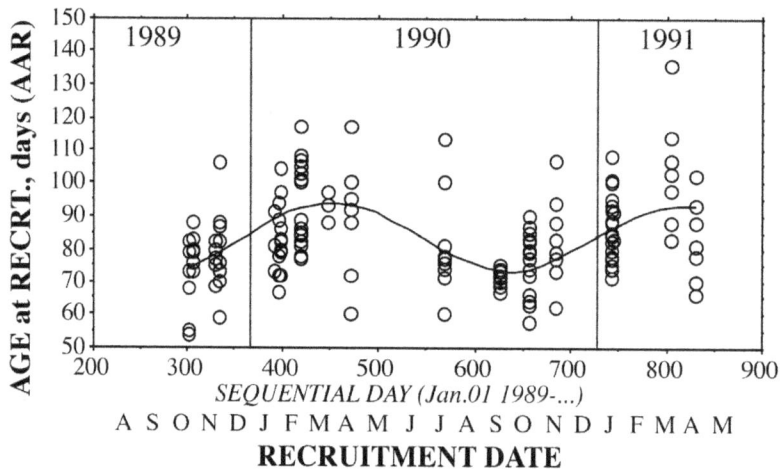

Figure 6-5. "Flat" plot: Y vs. cyclic X. This graph shows age-at-recruitment (AAR) data for *Sicydium punctatum* vs. date in linear format, with regression curve. Data as in Figure 6-6 and Figure 6-10. Data are graphed over more than one cycle, which is useful for inspection of the similarity of behaviour in successive cycles; the polar plot is less suitable for that. Few programs will produce the periodic function as a curve, even though they can calculate the regression. How to get the curve? Any regression line indicates where the fitted and predicted values lie, so even if your program won't draw the line, you can plot (and then trace through) these.

6.5.1. FLAT PLOT: Y VS. ANGULAR (CYCLIC) X

Periodic data can be plotted in a number of ways; most easily and perhaps most usefully as Y vs. X in cyclic or angular format (Figure 6-5). Here for example (Bell et al 1995) age-at-recruitment (AAR), meaning the age at which the fish (*Sicydium punctatum*) recruit from sea to fresh waters; these data show highly significant periodic relation to season. We plot AAR vs. DOY$_{recruitment}$. The regression line is drawn as a sinusoidal curve along with the raw data.

Several alternative plot types are conceptually related to this plot (Figure 6-5). Firstly, if it were rolled up into a tube (graph inside) having circumference = 1 period (365 days in this case), and we looked down through the top end, we'd see something like the polar plot (Figure 6-6). Or, secondly: if we printed the conventional (Figure 6-5) graph on clear plastic, rolled it up and rotated it while viewing through the roll (Figure 6-10), at a particular angle the data (if there is a nearly sinusoidal relationship of AAR and time) would form an elongated cluster with a positive slope (Figure 6-10); the rotation from the arbitrary zero to the acrophase would be ∂. From the other side $(0 + \partial + 180°)$ it would have a negative slope; from 90° off these angles the data would look like an ellipse (or a donut tilted directly out of or into the paper) and the correlation coefficient would be zero.

6.5.2. POLAR PLOT

The regression Eq. 6-3 or/and data may be also presented as a polar plot (Figure 6-6)—note that a polar plot doesn't require the polar angle system, and here we choose to use azimuthal references (Sec. 3.2).

The polar plot represents date (the x index) by an angle, and y by the radius to each data point. The angle of course has both sine and cosine components, used here as x'' and y'' coordinates (if we had plotted in the *polar angle* system which begins and proceeds anti-clockwise from the positive X axis, these components would be represented instead by cos and sin—but otherwise the plot would look identical). The regression line (where the fitted Ys are) is usually a shape that looks circular to ellipsoid but can be cardioid (or a twisted shape with negative values). The offset of the function indicates Y (AAR in the example) is greater in some directions (times of year) than others. The direction of offset indicates the time of year when Y is highest, and can be calculated by Eq. 6-9, Eq. 6-10, and Eq. 6-8 from the parameters (B_1, B_2) of Eq. 6-3, or approximated visually from the plot. Here, Y is the AAR (age-at-recruitment) of an *S. punctatum* sampled in Dominica, W.I. 1988-1991 (Bell et al 1995) on various dates X.

The polar plot here (Figure 6-6) incidentally demonstrates periodic regression's tolerance for non-uniform data density over the cycle; the presence of more points at some times of year does not 'pull' the phase angle in the same way it would 'pull' a mean vector (Sec. 6.4.4, Table 6-V). In fact, here, the acrophase or phase angle is opposite the location of highest data density. A higher density at any x in a regression merely improves the precision near that x.

How do you make a polar plot when there's no handy "polar plot" command in your software? The axis labels of Figure 6-6 show the method. It's really the reverse of scaling a vector to sit on the unit circle, as in Figure 4-4. We use the coordinates (x''_i, y''_i) on the unit circle that represent the original X (date or angle) of each observation, and multiply each by y_i (here AAR) to express it as a vector[f] or as polar coordinates.

So, create new X_{polar} and Y_{polar} columns such that $x_{polar,i} = y_i *$ $\sin`DOY_i$ and $y_{polar,i} = y_i * \cos`DOY_i$, then plot Y_{polar} vs. X_{polar}. Simple—it takes only minutes in a spreadsheet. Fitted values (often called Y-hat, conventionally $\hat{y}$, here also ¥) can be similarly treated to show the location of the regression line (here an ellipse); label and re-draw as needed. Now, as in Figure 6-6, the timing of each [positive] observation Y_i is indicated by its direction from the origin, and the value of Y_i is shown by its distance from the origin.

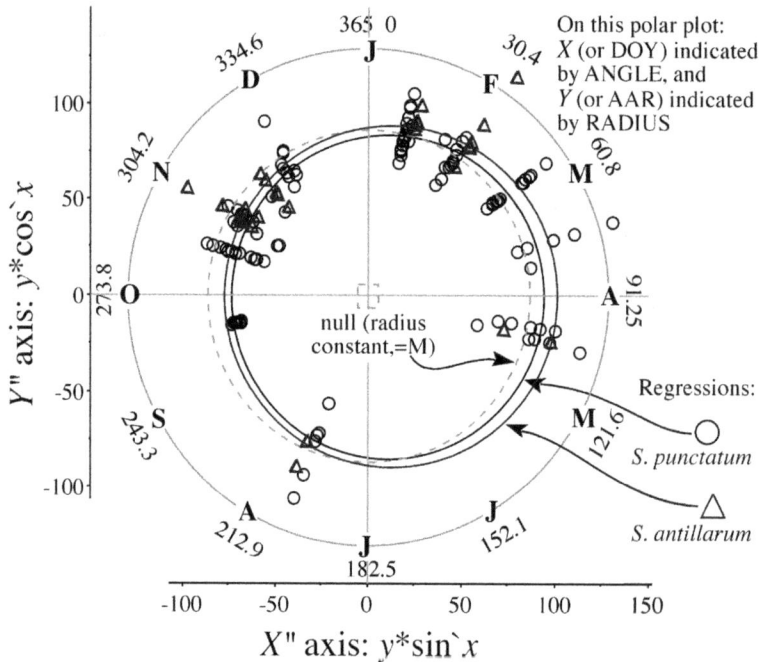

Figure 6-6. Polar plot; data and periodic regression, comparing two populations. X is represented by angles, and Y by distance from center. Data as in flat plot of Figure 6-5; axes are sine and cosine, scaled ($X_{polar}= X"=$ $Y*\sin`X$, and $Y_{polar}= Y"= Y*\cos`X$) to accommodate both age at recruitment (AAR) and timing. Axis labels show calculations that, plotted as $Y"$ vs. $X"$, create a polar plot. Centered dotted circle (radius = mesor = B_0) illustrates null hypothesis (Ho, rejected), which asserted that AAR = periodic mean or mesor at all angles (pairs of (sin,cos)), i.e. that B_1 and B_2 are zero. *S. antillarum* data were plotted slightly clockwise for visibility.

---

[f] Think of one single point i with magnitude y_i (or L_i) at a date (or angle); the polar plot lets you draw that angle through $(\sin`[date],\cos`[date])$, and make a line with length y_i; the terminal coordinates $(x_{polar,i}, y_{polar,i})$ of the vector for y_i at that date are thus $(y_i*\sin`[date], y_i*\cos`[date])$.

All the above has assumed that Y is never negative. Negative values pose a problem for polar plotting (if, that is, you want readers to understand them).

Negative values are tricky in polar plots

Many variables include negative values, so it's likely we will some day need to plot them. While negative data are readily displayed in an ordinary 'flat' plot of Y vs. X_{cyclic} (Figure 6-7), they require special treatment in a polar plot.

To illustrate, I used an arbitrary sinusoidal function, $Y = 8 + 20*\sin`X + 8*\cos`X$, to generate data with some negative Y values (Figure 6-7). On an uncorrected polar plot (Figure 6-8) the negative values cause the plot to reverse, loop, and reflect across the origin. Amplitude is 21.54; this exceeds the mesor by 13.54, which therefore is the minimum correction to eliminate the hard-to-read negative amplitudes. The correction applied was 30 (Figure 6-9).

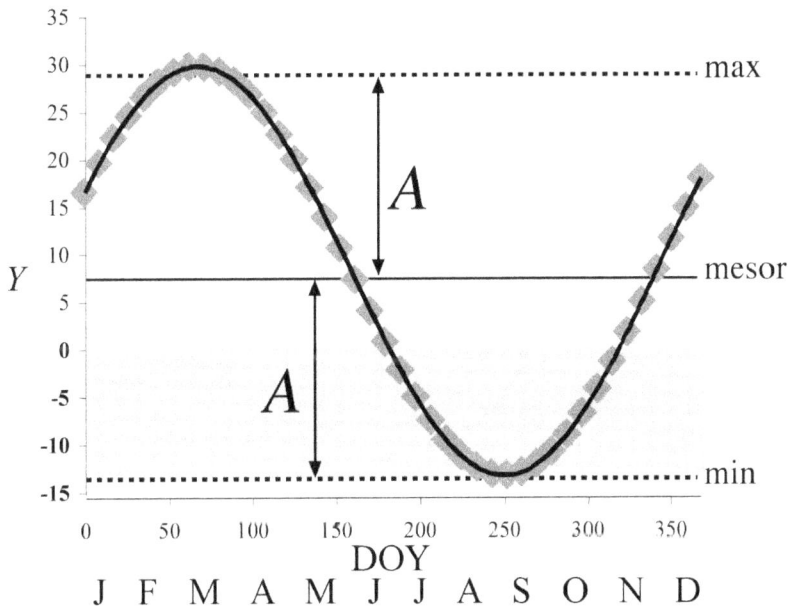

Figure 6-7. The conventional 'flat' plot: Y vs. X_{cyclic}. Example is a simple sine function $Y = 8 + 20*\sin`X + 8*\cos`X$,. Mesor, maximum and minimum are drawn onto the plot. Shading shows the troublesome region $Y<0$, where fitted Y 'reflects' across the origin in Figure 6-8.

On the uncorrected polar plot (Figure 6-8), the negative values lie not in the direction of their index (the value or angle x_i with which they are associated) *but in the opposite direction*. There is nothing intrinsically wrong with that as a convention but most readers will probably find it confusing.

We might well have guessed the sinusoidal function on a polar plot would be an ellipse; but it isn't. In fact, if Y is not always very positive the plot will look like an ellipse with a flattened portion, then (as the amplitude approaches the value of the mesor) cardioid. You can also get a lily-pad shape (very cardioid) if you polar-plot a function in which the amplitude is nearly as large as the periodic mean or mesor: the function will curl in, approach the origin, and then curl away again.

It gets worse: if Y becomes negative (i.e. amplitude>mesor), the negative values end up crossing the origin and opposite the angle to which they belong (Figure 6-8). That is because each point is in effect a vector with length L from the origin; the negative vector on an uncorrected graph is simply a vector running away from the angle it is associated with, instead of toward (as a positive vector does). However explainable it is, it remains tedious if not dizzying to read and explain in a paper. What we want is something intuitively readable. For that, we need all the values to lie in the direction of the index value (date etc. associated with the observation).

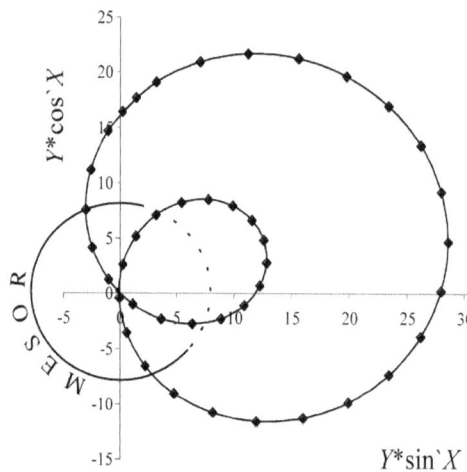

Figure 6-8. Uncorrected, nearly unreadable, polar plot of a function that is partly negative. Data as for Figure 6-7 and Figure 6-9. It is pretty but, to say the least, difficult to interpret. Negative values are 'reflected' across the origin, and thus align with dates opposite to the ones they are associated with. Because of that, superimposed scale indicators such as the mesor illustrated here *cannot relate to the reflected portion of the plot.* Likewise, a circle could readily be drawn to illustrate the function's maximum, but a circle to illustrate the minimum would relate only to reflected values. Therefore, such un-corrected polar plots, unless explained, assume very astute readers.

The solution is to plot with a correction added to all values (Y and Y_{fitted}), so that the negative values become positive. That will place a zero (imagine it as a circle) some distance out from the origin, with negative values inside the zero circle; you'll need to add a radial scale for Y.

The minimum correction needed to eliminate origin-crossing is equal to the amplitude minus the mesor (at that point the function looks like a lily leaf). The larger the correction used, the more towards circular the plot looks; that is not an error, it is simply that the human eye sees the radii from the origin to points on the function as being more uniform when they are all increased by the same amount; the difference in radii still exists. You can later edit the drawing. The axes now include your correction, so they look a little less like your original data, but to help your reader you can make a radial scale for value of Y. The radial scale replaces the scales that include your temporary correction (Figure 6-9).

Thus the purpose of your correction was simply to re-organise your polar plot and make it fit a radial scale that would be more readable than a curve reflecting across the origin. That saved you a troublesome explanation in your paper.

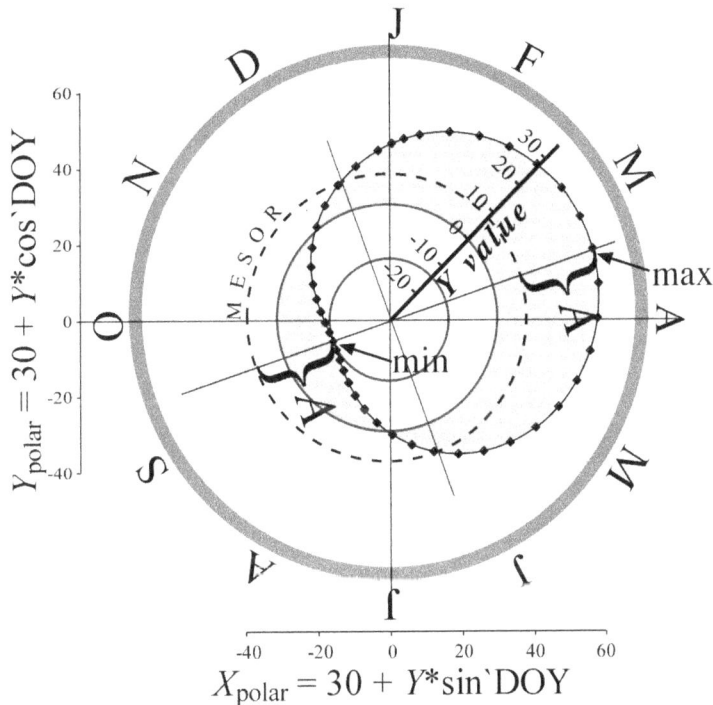

Figure 6-9. Polar plot modified for readability of negative Y values, using the same artificial data as Figure 6-7. The outer axes X_{polar} and Y_{polar} show how the data was plotted; they may be discarded once the other scales, labels, etc. have been added. The corrected zero is indicated by a reference circle crossing the radial scale at 0. The data region—between the function and the zero-reference circle—is shaded. The labels A (amplitude), max. and min. are not generally needed, but included here to help relate this plot to the key definitions. Are there too many points close together in that flattened portion nearest the origin? no; on an angular basis the density is the same as elsewhere—just count the points in each quadrant.

6.5.3. "Cylindrical" plot: *Y* vs. cosine of rotated *X*

The "cylindrical" plot of *Y* against a rotated & transformed index variable (Figure 6-10) is useful for teaching, and for convincing people who can't see the difference between a sine function and a straight line that there really is a relationship.

$$y = 10.102x + 83.835, r^2 = 0.281$$

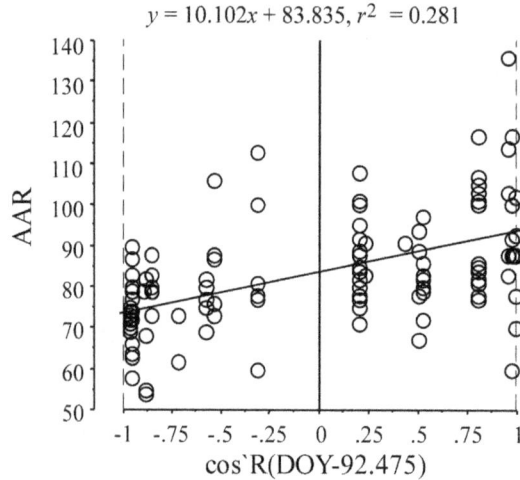

Figure 6-10. "Cylindrical" plot: linear regression *Y* vs. transformed rotated *X* (DOY). This is the same age-at-recruitment (AAR) data for *Sicydium punctatum* as in Figure 6-5 and Figure 6-6, but plotted as linear regression against a lagged and transformed *X*, in the form conceptualised in Figure 6-4. Seen as a straight-line regression it looks much more familiar. If you were successful in imagining Figure 6-6 to be like looking down the tube formed by rolling up Figure 6-5, then this figure is like looking sideways through that tube from the direction that brings the data into the clearest positive slope.

The linear or "cylindrical" plot could be calculated either of two ways: [A] as Eq. 6-3, AAR = B_0 + $B_1 \sin(R^{\cdot}(X))$ + $B_2 \cos(R^{\cdot}(X))$, then converted to the form of Eq. 6-5 or Eq. 6-4; or [B] as Eq. 6-4 iterative fitting of AAR = B_{0_n} + $B_{1_n} \cos(R^{\cdot}(X-\partial_n))$, where R denotes a radians transform and ∂_ns are the guessed phase angles between the peak of *Y* and the nominal zero of the cycle, and the best fitted ∂ is that giving the strongest positive correlation of *Y* vs. *X-∂*; resulting, for that example, in the regression *Y* = 83 + 10.1*cos([$2\pi/365$=0.0172142]*[DOY-92.475d]). Iterative fitting means preparing a number of regressions in this form using various choices for the lag ∂, and choosing the one that produces the best positive *r* (not r^2, see Sec. 6.4.2!). Fitting this angle is worth 1 DF, which is why the regression significance estimated after choosing this needs this DF as well as the one for the slope, and why the F-ratio and significance—but not the coefficients or r^2—will be over-estimated if this DF is neglected. In contrast, Eq. 6-3 converts the circular *X* into a (sin`*x*,cos`*x*) pair of derived linear *X*-variables, properly presenting the 2-dimensionality of a cycle to the regression routine.

6.6 MORE COMPLEX PERIODIC REGRESSION: MORE CYCLES, LINEAR TRENDS, CATEGORICAL VARIABLES

There is little difficulty modelling multiple cycles, and also other linear X variables, in the same regression.

In any complex model with linear and circular X variables, there's a possibility of correlation between some of the variables. We need to ensure that none of the linear X variables is already correlated with any of the cycles. That would be a likely problem if one wanted to model the effect of season as well as temperature on some Y (e.g. plant growth).

For example, temperature would likely be correlated with season and time of day, and that could lead to mis-attribution of seasonal variation in Y because temperature could act as, virtually, a proxy for season. To accurately dissect temperature from season we would need to de-trend temperature, i.e. remove the trends due to season and time of day, so we would have temperature anomalies. The first step in removing those trends would be—you guessed it—a periodic regression with temperature as the dependent variable (under Sec. 6.4.6: Intercorrelated Xs (e.g. de-trending temperature)).

For example, a single model could include effects, on Y (e.g. air temperature), of both time of year (DOY) and time of day (TOD):

Eq. 6-13: $Y = B_0 + B_1 sinR\grave{}DOY + B_2 cosR\grave{}DOY + B_3 sinR\grave{}TOD +$

            $B_4 cosR\grave{}TOD + error$

The residuals (errors) of that regression are Y (*temperature*) with the trends removed, so we could now use these as an independent (X) variable in a regression with the same cycles by which they were de-trended.

Multiple cycles are easily handled in a single regression, and can be obtained from a single index variable; e.g., calendar date and time can be converted to report many different cycles (yearly, lunar, daily, tidal). The proper conversion to radians is indicated (Sec. 13) by $R\grave{}(X)$; or to any standard angular units by $\grave{}X$. Conversion is described in Sec. 3.5 and illustrated in Table 6-II. For example, DOY with precision can give multiple cycles (Figure 5-1), e.g. both the time in the year (seasonal cycle) and the time in the day (daily cycle):

Eq. 6-14: $Y = B_0 + B_1 sin_{season}\grave{}DOY + B_2 cos_{season}\grave{}DOY +$

            $B_3 sin_{daily}\grave{}DOY + B_4 cos_{daily}\grave{}DOY + error$

which simply means:

Eq. 6-15: $Y =$ mesor + seasonal variation + daily variation +

            error (i.e. variation unexplained by this regression)

where, as before, the grave symbol $\grave{}$ is an operator indicating the proper angular transform to standard units for taking sine, cosine, etc.

The use of the same raw X (DOY) for two cycles—year and day—may seem counterintuitive, but think (Figure 5-1) of continuous time as a thread that can be wound onto spools whose circumferences represent different periods. Or, imagine a tractor, with the little front wheels representing a small cycle, and the large rear wheels representing a larger cycle, rolling over the same road that represents continuous time.

Here (Eq. 6-14), the conversion $_{season}$ = $2\pi/365$ to convert DOY (e.g. 209.85) to radian measure (e.g. 3.61rads) that reflects time of year; and $_{daily}$ = $2\pi/1.0$ to obtain radian measure for daily cycle. Really, for the daily cycle in this example, we want the sine and cosine of 0.85 cycles, but it's easy because $\sin`(x)$ = $\sin`(x+n$_whole_cycles$)$, so if we set k as 1 then $\sin`(0.85)=\sin`(209.85)$. DOY must of course have a proper decimal portion reflecting time in day.

Periodic regression can also include non-circular terms to reflect additional sources of variation. This is useful in case there is a longer-term trend superimposed on the periodic effects. E.g., if Y over a number of daily cycles includes longer-term effects, those may be accommodated by a linear term. This can also be important in multi-year studies, where there is, in addition to the cyclic patterns, a monotonic trend. To enable accounting for linear trends, a continuous time variable, e.g. elapsed time (ELAPSEDTIME), can be simply added, as in:

Eq. 6-16: $Y = B_0 + B_1\sin R`TOD + B_2\cos R`TOD +$

B_3ELAPSEDTIME.

A linear term can approximate short segments of a sinusoidal variation, so if there is very long-term sinusoidal variation then a linear term may often be useful and economical for data spanning only a small fraction of the longer cycle. Qualitative variation can also be included (nominal scale data) if your software permits use of a 'category' statement, or if you wish to construct your own dummy variables. That will give a coefficient for each category, but if you think the slopes, or responses to the cycle or other variables, vary by category then you should probably use separate regressions.

As with polynomial regression, avoid adding too many terms and over-fitting the data: a rule of thumb is that $n \geq 2m+1$. Select terms that have logical or theoretical support.

7 COMPARING TWO OR MORE INDEPENDENT POPULATIONS

Sometimes we need to compare the response of more than one population to a given cycle. That's different from comparing the relative importance of each cycle within one regression, which we can do, somewhat, by comparing amplitudes calculated from the regression coefficients for the sine and cosine components.

In this section we are interested in comparing the independent responses of different data to the same cycles. As usual, "cycle" can be temporal or directional.

For some situations, comparison can be done by pooling multiple populations of data for periodic regression, and then comparing (ANOVA etc.) the residuals amongst populations.

To analyse whether several populations have similar phase angle or direction, the Rayleigh test can be used with the phase angles obtained by periodic regression. It tests a sample of angles and determines whether or not there is an significant central tendency. There are dangers in applying it to average directions derived from sampling that is biased within a cycle, unless the average directions are those obtained from a process (like periodic regression) that is quite robust to non-uniform sample density over the cycle.

7.1 RAYLEIGH TEST: COMPARING PHASES OR ANGLES

The Rayleigh test assesses uniformity of a single, simple, list of directions, angles, times, or values of a circular variable. It can therefore deal with multiple units—and the units can be individuals or populations, provided there is only a single piece of angular information from each. That single piece of angular information can be a direction or an average of directions, or an estimated mean direction, etc.

It is a one-sample test to evaluate whether points (directions) are randomly scattered or show a significant central tendency (of direction, angle, timing). It can be used whether the data derive from individuals, in which case the test informs whether or not there is a significant departure from randomness; or derive from populations for which (e.g.) phase angles (locations where the effect of a periodic variable on some Y is maximised) have been estimated by periodic regression.

The data are simply a list of occurrences (Xs with no Ys, or Ys with no Xs) on a circular scale. H_0 (randomness) is rejected based on the length of the Mean Vector. The test assumes:

[1] that each observation is an angle, univariate: vector length (Sec. 4.1) is disregarded (which amounts to assuming all vector lengths are

equal at unity—if vectors length vary and are important then the Rayleigh test is inappropriate);

[2] that the unit circle has been sampled equally intensively (this is a *crucial* assumption) or that the measure is not biased by the sampling pattern (phase angles from periodic regression meet this requirement, because it is fairly robust against distribution of sampling effort).

The Rayleigh test treats data as univariate (it analyses only position in cycle), as angles, or directional vectors having length = 1, or points on a unit circle. It uses the logic that the mean vector's length (MVL, see Sec. 4.1, Figure 4-1, Figure 4-2) provides an index of similarity of constituent angles or directions, and can be used in a test. MVL equals the resultant vector length divided by n, and $0 \leq MVL \leq 1.0$. In resultant and mean vectors, opposing components have cancelled out. For example, if two directions on the unit circle (vectors of length 1) are nearly opposite, the MVL is nearly 0 because they cancel each other out. That near-zero MVL would indicate near-zero agreement in the directions. If directions are close, their MVL will be close to 1.0. That is the basis of the Rayleigh test.

The test calculates the length of the mean vector (Sec. 4.1). Theoretically, it divides it by the crude average of lengths of vectors—they're not the same thing! The test works because the crude average is always 1, by definition, and the $MVL \leq 1$, always, so the test statistic is technically a ratio, $0 \leq ratio \leq 1$; but in effect the ratio equals the MVL.

The test should not be applied to variables like sampling dates that are determined arbitrarily; a significant result would merely indicate bias of sample dates. So, if 'every day' means your summer field season, then you should not use the Rayleigh test because [a] practically, your results are biased by your absence over a significant part of the year (cycle), and [b] philosophically, the result would refer only to your activity, and would be meaningless regarding natural phenomenon not under your control.

But the test could be applied, say, to the first directions taken by animals after a certain stimulus, or the magnetic orientation of rocks, etc., because those are not affected by sampling choices. Or if observations throughout a cycle (without bias) recorded whether a particular event occurred or not, e.g. you have 365 days of observations on whether or not it snowed, then the dates of occurrence could be analysed using the Rayleigh test.

If on certain nonrandom dates a variable (*Y*) is measured such that the data are paired (date, variable) like (x,y), then periodic regression would be more suitable. Periodic regression tolerates bias in sampling time, because regression in general does not require data density to be even over the range of *X*.

In the context of evaluating whether phase angle (as indicated by regressions) is similar amongst taxa, each phase angle is a simple one-dimensional value, so a list is readily evaluated by the Rayleigh test.

Table H of Batschelet (1981) gives the probability of exceedance of mean vector lengths for various n and p (or alpha). You pick the significance level appropriate to your situation, look up the critical mean vector length for the n you have, and see if your mean vector length is greater than the critical value.

Example: comparing timing of migration

Seven taxa (4 spp. of fish, a complex of shrimps, and 2 spp. of snails) in Dominica, West Indies, are involved in this example.

Table 7-I. Rayleigh test for directedness of phase angle of peak abundance (as catch-per-unit-effort, CPUE†) amongst 7 taxa on seasonal and lunar scales. Data are one point per taxon, and the Mean Vector Length (MVL) if shown includes all taxa up to that row. **A**: Reference critical MVLs from Batschelet's (1981) Table H, for n=5, 6, and 7 for 3 levels of significance. **B**: For 7 taxa ("*S.*"=*Sicydium*, "*E.*"=*Eleotris*), their phase angles x_i obtained from regressions (elsewhere), in units of days after the arbitrary zero. Mean vectors calculated for 5 taxa, 6 taxa (includes all with a marine stage), and 7 taxa as shown. Significance indicated by asterisks as in part A.   (†Author's data)

A.	Critical MVLs (from Batschelet Table H)		
for	n=5	n=6	n=7
alpha ≤0.05*	0.76	0.7	0.66
alpha ≤0.01**	0.9	0.82	0.78
alpha ≤0.005***	na	0.88	0.82

B.		CPUE peaks ANNUAL (DOY)			CPUE peaks LUNAR			
for harmonic (k)		*1st (365d)*	*2nd (182.5d)*		*1st (29.5 d)*		*2nd (14.75 d)*	
arbitrary zero		Jan 0	Jan 0 & Jul 02		LQ (days after)		LQ & FQ	
n	Taxa in n	X_i	MVL(n)	X_i	MVL(n)	X_i	MVL(n)	X_i	MVL(n)
1	*S. punctatum*	331		24		10.7		8.1	
2	*S. antillarum*	322		45		11.3		8.6	
3	*E. pisonis*	339		9		8.2		7 2	
4	Gobiesocidae	271		85		11.9		7.1	
5	Decapoda	325	0.918**	117	0.295 ns	8.2	0.945**	7.8	0.972**
6	Neritidae	352	0.909***	127	0.102 ns	3.8	0.824***	3.7	0.801*
7	Thiaridae	41	0.815**	65	0.229 ns	25.0	0.584 ns	12.7	0.599 ns

Data are numbers per trap day (catch-per-unit-effort, CPUE) of fish, shrimps and snails caught in traps that primarily caught upstream-migrant stages (the only stages analysed for taxa 1-3, 5, and 6 (Table 7-I), although older and non-migrant individuals and taxa sometimes entered traps. All taxa except Thiaridae are juvenile-return anadromous (also called amphidromous), having marine post-larvae.

The phase angles are those obtained for each taxon by periodic regression. Periodic regression does not require a particular sampling regime; thus we can treat these phase angles like a list of occurrences (x)

un-constrained and un-biased by sampling regime, and suitable for the Rayleigh test.

Similarity of phase angles of recruitment amongst taxa and for first and second harmonics (Sec. 13.4) of seasonal and lunar cycles was evaluated by Rayleigh test on acrophases on four cycles, using reference critical values (Table 7-I-A) from Table H in Batschelet MVL, were calculated for each set of data to be tested. Here, MVLs are derived from phase angles, but generally any set of angles can be the source data.

The Rayleigh test results (Table 7-I) in row 6 are for taxa 1 to 6; results in row 7 are for row 1 to 7, etc. The phase angles of the second harmonics of the seasonal cycle showed no agreement amongst taxa. Amongst the 6 taxa having a marine postlarval stage, phase angles tend to be highly significantly ($p<0.01$) similar for the primary annual and lunar scales, and significant ($p<0.05$) for the secondary lunar scale, so randomness is rejected as an explanation for agreement on these circular time scales.

Having obtained a statistical result that tells us there is something going on here, we could proceed to answer the 'so what' question, putting the non-trite results in context and explaining implications. For this example, that could be, briefly: such agreement amongst very diverse taxa (fish, molluscs, and crustaceans) is remarkable (remarkable being substantially what a low p value means, unless the result is trite). Seasonality might have been unexpected in a tropical habitat, and therefore non-trite; but once noted the mechanism is obvious. Lunar synchrony (seen in the anadromous taxa) suggests tidal influence, but, although generally obvious, it can be eliminated on local physical oceanographic grounds. Logical interpretation would generally follow what was common to these taxa. Common elements are (at least) that the anadromous taxa in this example share both a habitat and an anadromous life cycle. A review (Bell 1999) indicated that, worldwide (i.e. not in the same habitat), gobies having this life cycle show a considerable range of phasing on the lunar cycle, so it seems that life cycle itself does not drive any particular phasing. From this example, migration cycle similarity in the same habitat seem a key ingredient of recruitment synchrony amongst diverse taxa. While there are few conclusive studies on predator swamping, it emerges by elimination and does have logical appeal as at least a benefit from the synchrony of recruitment in these taxa. We suggest this as a basis for the observed synchrony in a given habitat, and note that under this hypothesis the lunar cycle can be cue without being mechanism, allowing phasing to vary worldwide. This hypothesis implies that in other habitats suites of similarly anadromous taxa should show similar lunar synchrony.

8 IN CONCLUSION:

Tools exist to incorporate cyclic variation into analysis. It is therefore not necessary to attempt to exclude it from the data. In other words, you need not constrain observations to the same time-of-day, or stage-of-tide, time of year, phase of moon, etc. (note that if you are concerned with more than one source of periodicity there may be no realistic possibility of keeping sampling within the same stages of both cycles anyway). You may still choose to constrain observations when the question posed is narrow and well-defined. However, biology is dominated by cycles, so respect for them and competence in bringing cycles out of data will be both satisfying and useful.

8.1 JUDGEMENT ON THE STATISTICAL OUTPUT

No statistical procedure will do your thinking for you: all require you to apply judgement. Judgement incorporates awareness of the data, recognition of problems like undeclared but important sources of variation, and the assumptions of the statistical procedure used, etc. Scrutinise regression results.

Examine residuals for their distribution and for trends or pattern with respect to the untransformed index or time variable (e.g. DOY). Residuals, if not supplied by the program you use, can be calculated as the observations less the fitted values. Examine residuals for normality to check that regression assumptions have not been violated. If residuals are not approximately normally distributed, the suggestion is that the response of the Y to some X variable is non-linear, so that transformations of the Y should be investigated.

As with all analyses, satisfy yourself that the results make some sense: do the results violate theory, and if so are they strong enough to question it? ... i.e. can you find an error in your analysis?

Neither accept analyses uncritically, nor be paranoid (see Sec. 11.5).

9 APPENDIX I: EXAMPLES: SETTING UP, TRANSFORMING DATA, PERIODIC REGRESSION

These examples are supplementary to those in the text.

9.1 EXAMPLE: BRONCHITIS MORTALITY 1674-1799

Landers and Mouzas (1988) use historic medical records to obtain burials due to various diseases from 1670 to 1819, and explore the relation of changing seasonal profile in burials to longer-term shifts in predominance amongst diseases.

Landers and Mouzas use an index, the Monthly Burial Index (MBI), which is corrected for variation in the month length and standardised to a mean of 100. Here we examine just a part of that: the example of relative (corrected for month length) seasonal incidence of bronchitis mortalities 1674-1799.

The authors did not calculate these regressions, but nicely presented their data in tables on which I congratulate them. (Presentation of this analysis does not imply endorsement of the conclusion by the authors.)

We shall analyse the seasonal dynamics of the monthly burial index (MBI) attributed to bronchitis ($MBI_{bronchitis}$) by Landers and Mouzas. The regression we shall run is of the form

$$MBI_{bronchitis} = B_0 + B_1*\sin\grave{}DOY + B_2*\cos\grave{}DOY$$

and as you can see we have restricted it to the first harmonic only.

The data arrive ("given" in Table 9-I) with date in a common format (e.g. month and day, MM and DD). The common formats, despite that we use them routinely, are not mathematically proper. For analysis the time variable has to be converted to a mathematically proper circular form and then to its linear components, the sine and cosine. These steps are shown in Table 9-I.

Dates given in common format must be converted first to Day-of-Year (DOY, 0-365), a circular format (it seems linear because you can plot it but then 365=0 so it's actually not linear; and it is like an angle but not in standard units). DOY is then transformed to radians (angular), from which sine and cosine transforms are calculated. The grave ($\grave{}$) mark (as in sin$\grave{}$DOY) simply implies (Sec. 13, Notation) the proper transformations. Your program will expect to see radians and will calculate sines and cosines on that assumption, and rads=units*$2\pi/k$ for any cycle of length k units; so, for the first harmonic calculate as cos or sin(DOY*$2\pi/365$), and second harmonics as cos or sin(($2\pi/182.5$)*DOY) or equivalently sin(($2\pi/365$)*2*DOY). The analysis is given in Table 9-II.

The regression is highly significant (p = 2.44E-06 = 0.000002), and has a high adjusted (conservative) R-squared, 0.93. This data set is non-orthogonal (Sec. 6.4.6) as calculated on the basis of DOY, because months have unequal length, but would be orthogonal if analysed on the basis of months if assumed to be of equal length.

Table 9-I. Conversion of data (observed MBI for bronchitis) for analysis and plotting. X data in formats: common (as provided), cyclic, angular, and trigonometric. Sequence of transformations for an annual-cycle periodic regression: MM and DD are converted to DOY, then (multiplication by $2\pi/365$) to Radians (R), then to the paired transforms (sin(R), cos(R)) that permit analysis. The Y is unchanged unless there is a need for conventional transformations (log, root, etc.). Only one kind of plot is addressed here: Y vs. X_{cyclic}. Fitted values (Y_{fitted}) form the regression line, plot-able against DOY; note that it is often useful to calculate fitted values for a larger number of X values to obtain a smoother curve for the regression line. We could include the error term ε on the left side of the regression, but then the right side would be observed Y, not fitted Y.

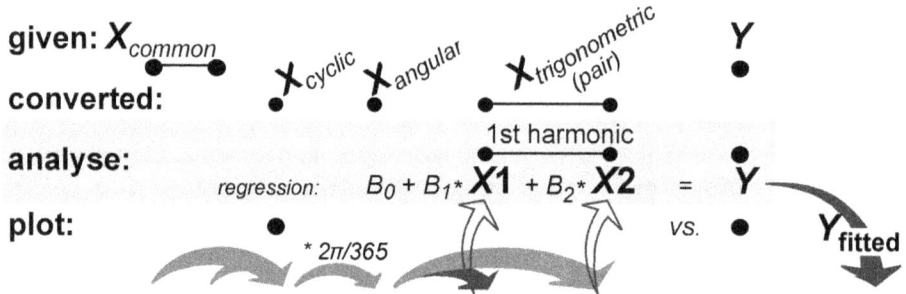

Month	MM	DD	DOY	R`DOY	sin`DOY	cos`DOY	MBI	MBI_FIT
Jan.	1	15	14.5	0.250	0.247	0.969	150	159.8
Feb.	2	15	45.5	0.783	0.706	0.709	137	150.7
Mar.	3	15	73.5	1.265	0.954	0.301	135	130.2
Apr.	4	15	104.5	1.799	0.974	-0.226	99	99.7
May	5	15	134.5	2.315	0.735	-0.678	72	70.2
June	6	15	165.5	2.849	0.288	-0.957	56	47.9
July	7	15	195.5	3.365	-0.222	-0.975	45	40.1
Aug.	8	15	226.5	3.899	-0.687	-0.727	42	48.4
Sept.	9	15	257.5	4.433	-0.961	-0.276	57	71.1
Oct.	10	15	287.5	4.949	-0.972	0.234	94	100.7
Nov.	11	15	318.5	5.483	-0.718	0.696	154	131.1
Dec.	12	15	348.5	5.999	-0.280	0.960	161	152.3

The periodic mean or mesor is 99.95, which in this example is uninformative about the population because it (100) is the value to which MBI was standardised in the data.

The amplitude, always in units of Y of course (here the monthly burial index), is calculated by Pythagoras' theorem from the sine and cosine coefficients.

The phase angle, the value of X at which the contribution of the cycle to Y is maximised (and here the cycle is the sole variable so it is also where the entire function is maximised), is calculated also from the sine and cosine coefficients, as shown in Sec. 4.3.

Table 9-II. Regression output for data in Table 9-I. The regression function is directly readable from the coefficients as: $MBI_{bronchitis}$ = 99.948 +13.28*sin`DOY +58.36*cos`DOY. Implied phase angles (dates) and amplitudes are calculated from the sine and cosine coefficients, and visualised in Figure 9-2. Second harmonic has (by definition) 2 peaks in one main cycle, 0.5 cycles apart; second peak shown in italics.

Regression Statistics

Multiple R	0.97127632		
R Square	0.94337768		
Adj'd R Sq	0.93079495	Amplitude:	59.9
Std Error	11.957238	Phase DOY:	13.0
Obs'ns	12	Phase Date:	jan13

ANOVA TABLE

	DF	SS	MS	F	p[F]
Regression	2	21438.9	10719.4	75.0	2.44E-06
Residual	9	1286.8	143.0		
Total	11	22725.7			

	Coefficients	Std Error	t Stat	P-value	Lower 95%	Upper 95%
Intercept	99.95	3.45	28.95	3.41E-10	92.14	107.76
sin`DOY	13.28	4.87	2.73	0.023	2.26	24.30
cos`DOY	58.36	4.89	11.93	8.09E-07	47.30	69.43

The plot (Figure 9-1) of the periodic regression function based on the first harmonic alone gives only a little evidence that additional harmonics are needed. Nevertheless, the second harmonic is often significant in real data, and often useful (markedly increases R^2).

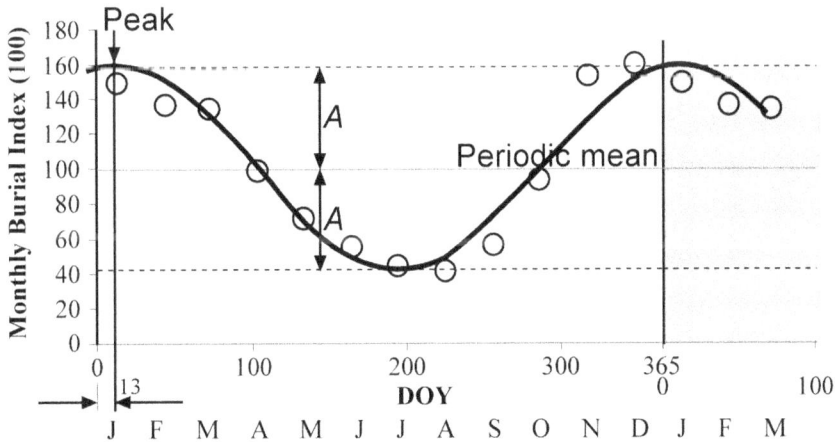

Figure 9-1. Bronchitis monthly burial index, 1674-1799: data (circles), fitted line (solid curve), periodic mean or mesor (M≈100), phase angle (peak location), amplitude (*A*), and limits (M±*A*). Shading marks repeated data (in plot, not analysis).

The phase angle can also be visualised (Figure 9-2) by plotting the sine vs. cosine coefficient, i.e. treating the coefficients as the terminus of a vector originating at (0,0).

As explained by the authors, bronchitis is a cold-weather disease, thus explaining why it peaks in the winter. Contemporary data can be used in models to inform about or predict the timing of need for health services, thus contributing to better deployment of resources. Changes in phase angle might be used to explore the health implications of changes in climate, or in house construction methods, or in cost and availability of heating fuel, clothing, etc.

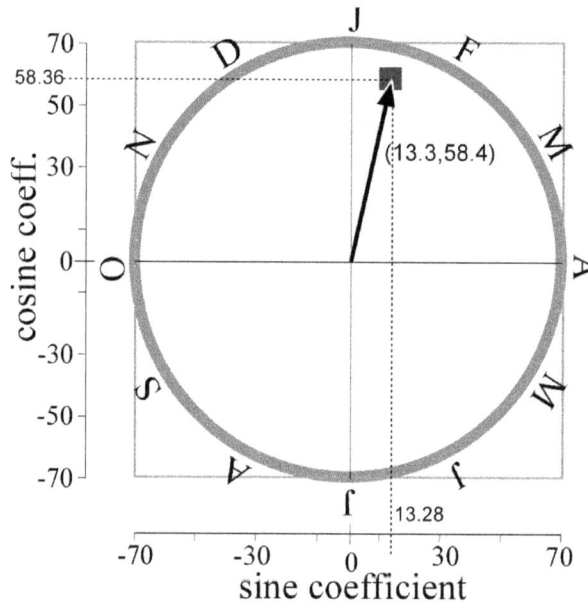

Figure 9-2. Visualising the phase angle from the sine and cosine regression coefficients. Here, from Table 9-II, the (sin,cos) coefficients are plotted to show the phase angle (direction of arrow) and peak height (amplitude, by length of arrow) above the periodic mean. (The year starts at the top of the circle because we are using the azimuthal system, Sec. 3.2)

9.2   EXAMPLE: URINARY NOREPINEPHRINE

Batschelet (1981) presents (his Table 8.2.1) data from Descovich et al (*1974; Age and catecholamine rhythms. Chronobiologica 1:163-171* (as cited in Batschelet)) on urinary norepinephrine. The task is to characterise the temporal structure of urinary norepinephrine in healthy human subjects. Here we adapt the data for spreadsheet and periodic regression, with months converted to radians. As analysed, the data set is orthogonal (Sec. 6.4.6), but if the varying length of months were acknowledged it would not be. Figure 9-3 shows what the data look like, with the fitted periodic regression line (drawn through fitted Y values).

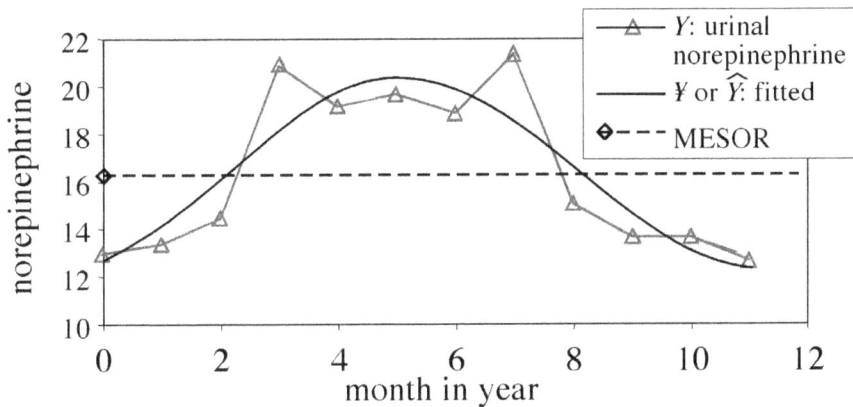

Figure 9-3. Annual cycle of urinary norepinephrine, Descovich data cited in Batschelet (1981) Table 8.2.1. Data (triangles) are shown with periodic regression (first harmonic only) as a smooth line, and mesor indicated by a diamond on the Y-axis. Month 12 = month 0 (nature of a cycle). It is fairly common to repeat some data to plot a little more than one cycle, to show the pattern more clearly; but the argument against doing that is it tends to support an illusion of more data than really exist—so, if you're planning to do that, you might consider greying the points or shading a portion of the graph and labelling it 'repeated', etc.

The periodic regression is run as a multiple regression (but various cautions apply, Sec. 6.4.6) in which one or more pairs of variables are the sine and cosine transformations of a circular variable. (Don't be alarmed if the spreadsheet shows tiny numbers like 2.46596E-12 instead of zeros; I'm told the problem is fundamental to binary math because of the difficulty in dividing 1 by any number, e.g. 3, which results in irrational numbers that cannot be exactly represented by a finite number of digits.)

This analysis is done to follow Batschelet's, although using the midpoints of months would normally be a better choice, and it would of course be far better to use the actual dates rather than monthly averages.

Table 9-III shows the data and the transformations leading to the regression. For comparison, or if you don't find this clear enough, see also Table 6-II for the sequence of transformations.

The final form of multiple regression submitted to software is: $Y = B_0 + B_1 sinR`X + B_2 cosR`X$; i.e. select Y as dependent and $sinR`X$ and $cosR`X$ as independent variables. Be sure your stats program is set to provide an intercept (not a regression forced through the origin); this is usually the default setting, but command-line programs may require you to explicitly include B_0 as "CONSTANT" or some such name in the model statement. The intercept will be considered a *mesor* because of its special meaning due to the absence of a true zero on a circular scale. The residuals should be examined for trends, and an assumption of the ANOVA associated with the regression is that the residuals be normally distributed.

Table 9-III. Annual cycle of urinary norepinephrine, Descovich data cited in Batschelet (1981) Table 8.2.1. (Alternative presentation: Figure 11-2.)

X (original)			Y	R`X	sinR`X	cosR`X	¥
Time variable (X): month in year. NOT usable in this form, needs transforming to sin & cos	Batsch-elet worked in degrees, not rads		Dependent variable; no transform-ation used	Trans-formed time variable R`X in units RADS	Proxy variable sin of angular transform of time	Proxy variable cos of angular transform of time	Fitted values (obtained using regression) needed to plot regression line
Mth	Mth #	Month as deg.	Y: urinary norepin-ephrine	RADS= Mth# *2π/12	sin (RADS)	cos (RADS)	¥
J	0	0	12.9	0	0	1	12.593
F	1	30	13.3	.5236	.5	.866	13.995
M	2	60	14.4	1.0472	.866	.5	16.001
A	3	90	20.8	1.5708	1	0	18.074
M	4	120	19.1	2.0944	.866	-.5	19.658
J	5	150	19.6	2.6179	.5	-.866	20.329
J	6	180	18.8	3.1416	0	-1	19.907
A	7	210	21.3	3.6652	-.5	-.866	18.505
S	8	240	15.0	4.1888	-.866	-.5	16.499
O	9	270	13.6	4.7124	-1	0	14.426
N	10	300	13.6	5.2360	-.866	.5	12.842
D	11	330	12.6	5.7596	-.5	.866	12.171

The tabular regression output (Table 9-IV) gives the equation:

$Y_i = B_0 + B_1 sinR`X_i + B_2 cosR`X_i + e_i$, or

$Y = 16.25 + 1.824 sinR`X + -3.657 cosR`X + e_i$

where Y_i = observed Y; e = error or residual; Y or ¥ = fitted or predicted Y. The lag or phase angle ∂ = arctan(B_1/B_2) + Quadrant correction (for (sin,cos) having signs (+,–), QC = 180°):

∂ = arctan$_{deg}$(1.824/-3.657) + 180° = 153.49138° (Batschelet gives 153.5°)

The resulting phase angle of 153° corresponds to DOY 153*365/360=155.13, which is June 05 (via DOY table Sec. 16).

Table 9-IV. Analysis output for urinary norepinephrine. Format is typical of many stats programs. Data are in Table 9-III.

Count:	R	R^2	Adj. R^2	RMS Resid
12	0.898	0.806	0.763	1.638

Analysis of Variance Table

Source	DF	Sum Squares	Mean Square	F-test:
REGRESSION	2	100.19	50.095	18.677
RESIDUAL	9	24.14	2.682	p = .0006
TOTAL	11	124.33		

Beta Coefficient Table

Variable:	Coeff	Std.Err.	Std.Coeff.	t-Value	p
INTERCEPT	16.25				
sin(RADS	1.824	.669	.401	2.728	.0233
cos(RADS	-3.657	.669	-.803	5.469	.0004

Amplitude = $(B_1^2+B_2^2)^{1/2}$ = $(1.824^2+(-3.657)^2)^{1/2}$ = 4.0866 (Batschelet gives 4.1 of whatever "special units" norepinephrine was measured in).

The mesor, or periodic mean value of Y, is 16.25.

Batschelet then rewrites the regression in 'cosine regression' format (vs. standard format, Sec. 6.4.1), as

{Y = mesor + Amplitude*cos(angular measure of months - ∂)}, or

general: $Y = M + A \cos`(t_i-\partial)$

radians: $Y = M + A \cos\{(2\pi/k)t_i - \partial \text{rads}\}$

degrees: $Y = 16.25 + 4.08*\cos`(t_i°-153.49°)$

noting that ∂ must in the same units as any term it is added to.

Most programs will not print a periodic regression line. You will need to plot fitted values (¥) to show where the line is. Plot Y_i and $¥_i$ vs. X in a linear or angular format. If a smooth curve is desired for illustrating the regression line, import resulting graph to a graphics program and trace a curve through the fitted values.

Note that although our regression was calculated using radians, the sines and cosines would have been the same whichever standard angular measurement unit was used for initial calculation. (We just have to ensure that the calculator is set to expect the units we give it.) Only when we set "deg" or "rad" when taking the $\tan^{-1}$ (to get ∂') did the result come in a particular system (deg, rad, etc.).

10 Appendix II: Spreadsheet Formulas & Macros

Much of data preparation & regression interpretation can be automated, either by using spreadsheet formulas and structure, or by using macros written in VBA (Microsoft's Visual Basic for Applications, which comes with Excel or Office).

10.1 SPREADSHEET FORMULAS

Looking up of DOY from a table for each entry when creating a data set, and in analysis the interpretation of phase angle and amplitude etc., is time-consuming if many regressions are to be run. Spreadsheets can be set up to do much of this. (With any program or spreadsheet set up to do statistics, it is important to check the results against known examples.)

10.1.1. GET DOY from MONTH & DAY

We often have a date, such as Sept. 24th, which we want to express as DOY. Excel (for example) offers no clean way to do this out of the box. While the Day-of-Year chart (Sec. 16) at the end of this book is useful for small tasks, conversion of large numbers of dates in a data set can be automated. Begin with your date in *yyyymmdd.hhmm* format (Sec. 5).

Macros are the best way to do these kinds of things, but you can do it with spreadsheet formulas. It's worth noting that spreadsheet formulas nevertheless have one advantage over macros: they travel with the spreadsheet even if it is transferred to a computer that doesn't have the same macros (in Add-ins) installed, so there will be situations where spreadsheet formulas are preferred.

Table 10-I. Spreadsheet formula converts month and day to DOY

statement for getting DOY from mm and dd:		
" =CHOOSE([mm], 0,31,59,90,120,151,181,212,243,273,304,334)+[dd]-1 *that is, for month x choose element x (9) in the list (of DOY values for the first of each month), add the date in the month, and subtract 1 (so the first day begins at mathematically proper 0)*		
input (e.g. Sept 24)	mm	dd
	9	24
DOY result:	266	

A simple spreadsheet formula (Table 10-I) to convert month and day to DOY uses the CHOOSE function; your help files will explain how the function works. The formula is a bit long, but it's nevertheless easy to copy it down an entire column if you wish and so it can process quantities of dates quickly. It uses a list of the DOY (0-365) values for the first day of each month, and then adds the day of month, then subtracts

1 to correct for the mathematical impropriety in our common format for dates (it is mathematically wrong for the first day of the month to begin at "1"; see Sec. 5).

Spreadsheet formulas can accept the date as *yyyy* and *mmdd* (e.g. "0802" for Aug 02), and then split the latter up automatically as *mm* and *dd*; and can then generate DOY and nDOY (Table 10-II). The same approach can accommodate a date put in as *yyyymmdd.hhmm*. I've included the **nDOY** column because it's so often useful, either in including the long-term trend in the analysis, and for plotting (nDOY always has a zero date, here Jan 01, 1993, which you can see from the formula). The sin and cos formulas are given there also; they require DOY (or nDOY).

Table 10-II. Spreadsheet formulas convert common formats to DOY, nDOY, sin`DOY, cos`DOY.

Ordinary year ... not leap-year (see below for that):
The formula for DOY (ordinary =CHOOSE(C15,0,31,59,90,120,151,181,212,
year) is: 243,273,304,334)+D15-1

	B	C	D	E	F	G	H	I
	yyyy	mmdd	mm	dd	**DOY**	*nDOY*	**sin`DOY** sin(inYR)	**cos`DOY** cos(inYR)
24								
25	*1993*	0324	3	24	82	82	0.9873	0.1586
26	*1999*	0508	5	8	127	2317	0.8165	-0.5773
27	*1993*	0509	5	9	128	128	0.8065	-0.5913
28	*2000*	0527	5	27	146	2701	0.5878	-0.8090
29	*1993*	0603	6	3	153	153	0.4863	-0.8738

Formulas underlying row 25 and subsequent. Formulas are staggered into other rows for readability)

25	1993	0603	=INT(C25/100)
25			=C25-100*D25
25			=CHOOSE(D25,0,31,59,90,120,151,181,2 12,243,273,304,334)+E25-1
25			=F25+(B25-1993)*365 Note: the 'start' year for nDOY is set at 1993.
25			
25			=SIN(2*PI()*F25/365)
25			=COS(2*PI()* F25/365)

You can make formulas leap-year clean if you want to, but it'll rarely be worth the trouble†.

(†because what you are really after with DOY as an angle is the position of the Earth in its solar orbit, i.e. the angle between one line from the sun to the earth and one from the sun to some distant star in the plane of the orbit of the earth around the sun, AND that being so the leap year is simply an intermittent integer correction of that, amounting to ca. 1 day out of 4*365, or 1/1460, and in angular deviation from 'true' it is only 0.5/1460 because that one day spans the greatest deviation above plus the greatest deviation below. Convinced?)

But, if you insist, use an IF statement conditional on year÷4 (and note that this does not take care of all leap years, just most of the recent ones) having remainder zero plus whatever other determiners of leap year you want to include, and based on that use a normal-year CHOOSE list or a leap-year CHOOSE list. It makes for a rather long formula:

"IF(MOD([year],4)=0,CHOOSE([Month1–12],0,31,60,91,121,152,182, 213,244,274,305,335)+[DayOfMonth]–1,CHOOSE([Month1–12],0,31, 59,90,120,151,181,212,243,273,304,334)+[DayOfMonth]–1)

The things in square brackets there are variable names, not expressions. And if you wish you can use an IF statement to make the denominator of R˙ leap-year clean, so that R˙ for leap years is $2\pi/366$. You would only take the trouble to be leap-year clean in very special cases. The error by not doing so is very small, much smaller (0.25 in 365) than errors (~1 in 24) accepted by previous authorities using very brief tables for converting to *sin* and *cos*.

10.1.2. GET MONTH & DAY FROM DOY

Suppose your regression coefficients imply that the phase angle (acrophase) of a cycle is DOY=273.56, and you want to express this in common format, you could use Table 16-I or a formula.

Table 10-III. Getting back common date (month, day) from DOY. *Formulas are copied as text to the right of the cells they are in.* The absolute references ($column$row) refer to the lookup portion of the table. [[a]Month No. in table is common, i.e. 1-12]

	C	D	E	F	G	H
102		*LOOKUP table*			input DOY⇒	273.56
103		*for DOY-to-Date:*				
104	DOY	Month name	Month No[a]			
105	0	Jan	1		TO GET:	RESULT:	FORMULAS LOOK LIKE:
106	31	Feb	2		DOY of 1stofmonth	273	"=LOOKUP(H102,{0,31,59,90, 120,151,181,212,243,273,304, 334})
107	59	Mar	3		month#	10	"=LOOKUP(H102,C105: C116,E105:E116)
108	90	Apr	4		day.dec	1.56	"=H102-G106+1
109	120	May	5		monthname	Oct	"=LOOKUP(H102,C105: C116,D105:D116)
110	151	Jun	6		day	1.00	"=INT(H102-G106+1)
111	181	Jul	7		part day	0.560	"=H102-G106+1-INT(H102-G106+1)
112	212	Aug	8		hour.dec	13.440	"=G111*24
113	243	Sep	9		mins past hr	26.400	"=(G112-INT(G112))*60
114	273	Oct	10		*TIME OF DAY:*		
115	304	Nov	11		hhmm.decmin	1326.400	"=INT(G112)*100+G113
116	334	Dec	12				

This conversion is a little more difficult than getting DOY from date. It needs more space because it needs a lookup table. The lookup table need not, of course, be near the cells that call it—you can place the LOOKUP portion of the table in one place and use any of the formulas anywhere else. (Recall that the $ sign preceding a cell's row or column reference makes it an absolute reference—see your spreadsheet help files to learn about absolute and relative references—we could have used these so that when the formula is copied to more cells it can still refer to the same $cells, instead of to the cell that bears the same relative position to the cell where you paste the formula.)

Don't let the potential for precision create an illusion of accuracy: times of day are unlikely to have much meaning on the annual scale, but the point is that with accurate date and time you can extract the phase of any observation even on small cycles.

10.1.3. GET PHASE ANGLE FROM SINE & COSINE COEFFICIENTS

Estimating the **phase angle** of a cycle in a periodic regression (Sec. 6.4.4) is like finding an angle from a sine and cosine, or from the terminal coordinates of a vector (Sec. 4.3). To visualise phase angle, see Figure 9-2.

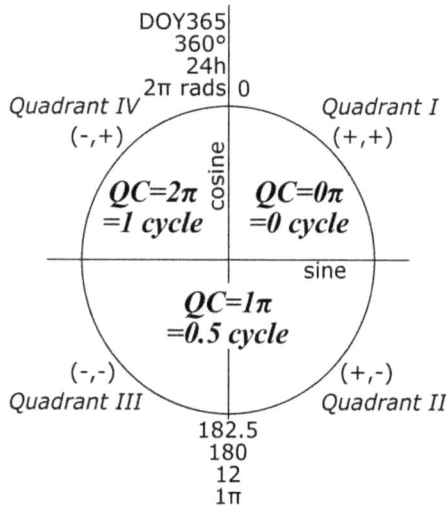

Figure 10-1. Quadrant corrections (QC) for correcting arctan(sin/cos) to true angle. Corrections to ∂', when obtaining $\partial'=\arctan(\sin\partial/\cos\partial)$, depend on the quadrant in which ∂ lies; quadrant is diagnosed from the signs of sin and cos (or the signs of the regression coefficients relating to sin and cos of a given cycle). $(0,1,2)\pi$ radians equate to 0, 0.5, & 1 cycle as general QC in any system. (See Sec. 4.3—remember that there is a true ∂, and the sin and cos coefficients refer to the *true* ∂, not ∂'; but the problem is that arcsin(sin∂) etc. are ambiguous.)

Given a sine and cosine pair or (x'',y'') coordinates of a vector, the tangent is sin/cos in either azimuthal or polar angle visualisations, or (azimuthal) x''/y'', or (polar) y''/x''. Arctan(sin/cos) gives an uncorrected estimate of the true angle. Addition of the appropriate quadrant correction (QC, Sec. 4.3.2) then yields the corrected or true angle. A spreadsheet formula giving the appropriate QC in radians is:

Eq. 10-1   =2*pi()*IF(COS<0,0.5,IF(SIN>=0,0,1))

An easy error is to get the wrong QC. Therefore, check and mentally evaluate the quadrant that is implied by the coefficients' signs. (Just as for an angle, you can get the quadrant mentally from the signs of (sin,cos) coefficients (parameters), and their relative size will tell you what part of the quadrant it is in.) Whichever method you use to find the QC, be alert to the possibility that it might not properly handle cases where the true angle lies exactly on one of the axes (either sine or cosine = 0).

The phase angle is then arctan(x''/y'')+QC; phase can be expressed in any cyclic units, whether radians, degrees, a decimal portion of a cycle, or units of the cycle itself (hours in day, days in year, etc).

The **amplitude** of a cycle is simply $\sqrt{[(sin\_coeff)^2+(cos\_coeff)^2]}$.

The **mesor** is exactly the intercept, though noting its special meaning as a periodic mean.

10.2 MACRO FUNCTIONS

Analysing cycles involves many conversions, most of which can be automated by macros written in Excel's VBA (Visual Basic for Applications). Once written, macros save a lot of time. Macros are of two kinds: functions and subroutines. What's the difference? A function is something invoked in a formula; it takes some information and returns something derived from it. A subroutine is something that can do more than that, e.g.: sort a list of values in a spreadsheet, run a simulation, or repeat an iterative procedure until some target is satisfied. Subroutines can be invoked by a button on a spreadsheet, or by name, or called by other macros. A VBA "Module" is a container, a subdocument of a spreadsheet file, for one or more macros.

Functions are our main focus right here. The VBA functions you write can be used in spreadsheets just the way we use Excel's built-in functions. Get to the VBA editor via Tools/Macros/ and Edit. You can also "Record" a macro, and then see it in the VBA editor.

Using the Functions

Use a macro function just as you would any other function. Simply type in a cell an equals sign, the name of your function, and the arguments in brackets. For example, the function ksin asks for k as well as x, allowing it to convert x from the scale k to $k=2\pi$. To obtain the proper sine of x=August 9th (DOY 220), ksin asks also for k to convert x to radians:

=ksin(365,220) [equivalent would be: =sin(220*2*pi()/365)]

and it returns –0.6016241, the proper sine of DOY 220.

As with any Function, if you aren't sure of the syntax (number and order of arguments, etc.), use Excel's 'Paste Function' button (looks like "fx"); your macros will be easily located under "user functions" or alphabetically (they all begin with the same letter), and follow the prompts. The prompts (from the Paste Function dialog) are more useful if the code used meaningful names for arguments.

Test the functions with a small list of angles that span a cycle. Take the proper sines and cosines, then use kPhase to recover the original angles, then compare. Values like 2.345E-17 where zeros ought to be merely reflect the limitations arising from binary math in the software.

Occasionally, you'll have a spreadsheet that has been opened where the Functions were at first unavailable, and, even after the Function becomes available, cells retain the "#NAME?" error. If you open a cell, place the cursor in it, and click enter, that cell will then show its result instead of the NAME error; but "kicking" all the cells this way is tedious. But to kick the entire worksheet, use the "Calculate Now" button; if it's absent from the setup where you're working, here's a silly dodge: use Edit/Replace to find "=" and replace with "=" (this amounts to no change, but it causes the formulas to look again for the functions they use).

Hints in writing macros:

Use meaningful variable names in the Function definition (first line), so that the Formula Insert button brings up meaningful prompts. If your function is defined "Function kmagic(r,h,s)" you'll never figure it out; define it using meaningful variable names, like "Function kmagic(rabbits,hats,sticks)".

Include comments in your code so you can understand it later. Comments are isolated by a single *straight*† quote, with the material following it considered a comment; comments can follow active code or can be on their own lines. († A curly or "smart" quote mark, which you can get if you paste from Word, will not be recognised in VBA, and will need to be corrected to straight.)

The single straight quote mark tells the compiler to skip the material to the right on the same line; it is thus useful while writing or testing code because it allows code to be made inactive without removing it. Instead of a quote to isolate comments, some other programming languages used the letters REM, for 'remark', and the use of REM to make codes inactive was called "remming out".

It's helpful to break a task up into discrete subtasks, and write each as a separate macro; one macro can invoke another, allowing the subtasks to be brought together. At the same time, this division reduces repetition because a component macro can be used (called, invoked) by any other macro.

How to put these macros into your own Excel macro set

Open VBA (via Tools/Macro/Visual Basic Editor). Select the document ("project") you wish to work in, and Insert/New Module. Begin each module with the statement "Option Explicit". This protects against variables retaining values between successive invocations in macros, and saves you having to set each to zero.

Type the Macros directly into a new VBA module, or into Word (plain text, straight quotes selected under Tools/Autocorrect/"Autoformat as you type") then copy and paste into a VBA module. You can get to the VBA editor via TOOLS/MACROS/Edit.

Either save into a new document that you will later save as an Add-In (you can then load the add-in via Tools/Add-Ins to make it globally available), or save into your "Personal Macro Workbook" etc. Either way will make the macros available to any Excel document on your machine (or account). Note: a spreadsheet won't carry the macros around to other computers unless you copy the macros into a Module in that spreadsheet.

The Add-In folder is usually in the Office folder. Locate it. Regularly duplicate your created Add-Ins; that protects against file corruption. Then, make copies to backup media to protect against your computer being stolen or the drive crashing. Then, put backups in a remote location to protect against meteorites striking your office. Etc.

Copy your macros from VBA to a text document, so you'll be able to adapt them when some other system replaces VBA. VBA is only available in some versions of Excel, and may be changed to another system in a few years. The logical flow of the macros will still be useful when translating to another system.

10.2.1. CORE MACRO FUNCTIONS FOR ANALYSING CYCLES

The code in boxes is the core of a set of macros that speed conditioning of circular data for analysis, and interpretation of periodic regression. Italics are to ease your reading; they will (should) vanish if entering into VBA editor in Excel.

In the functions, I've put cycle length (the period) first, to simplify conversion to spreadsheet formulas (useful if sending to a colleague without the same macros). To convert from macros to Excel formulas, run a find-and-replace so that every "ksin(365," (including the comma!) would be replaced by "sin(($2\pi/365$)*", etc. (the brackets must be incomplete because only the front, constant, parts of expressions are changed). Even that find-and-replace can be recorded as a macro ....

Data prep: proper sine and cosine:

Sine and cosine are the key for analysis and calculation. Most programs require angles in radians, and standard functions like "sine(x)" require conversion of x to radians first. That's how you get the *proper* sine etc., which I indicate by the grave mark as in "sin`x" (Sec. 13.2). My sine and cosine functions convert automatically, using any cycle length (period) you specify. The standard Excel function is, in effect, sine($2\pi,x$); enjoy this confirmation that, whether explicit or implicit, an angle has two dimensions—either period and x, or sine and cosine.

Task: SINE of anything [ksin(cycle length, units)]

```
Option explicit 'begin any module with this line, for safety
' Copyright K.N.I. Bell (this notice to be kept with any or all of the macros
        here)
Function ksin(cyclelen, x) 'cyclelen is k
'returns proper SIN of x for any period or cyclelen (same units as x)
If cyclelen = 0 Then cyclelen = 2 * WorksheetFunction.Pi()
' 0 allowed as a shorthand code for rads, saves typing value of "2*Pi()"
ksin = sin(x * 2 * WorksheetFunction.Pi() / cyclelen)
End Function
```

Task: COSINE of anything [kcos(cycle length, units)]

```
Function kcos(cyclelen, x)
'returns proper COS of x for any period or cyclelen (same units as x)
If cyclelen = 0 Then cyclelen = 2 * WorksheetFunction.Pi()
' 0 allowed as a shorthand code for rads, saves typing value of "2*Pi()"
kcos = cos(x * 2 * WorksheetFunction.Pi() / cyclelen)
End Function
```

For example, the proper sine of the 17th hour of the day on the 0–24 scale would be returned if you typed the formula "ksin(24,17)". It saves a step (17 * $2\pi/24$) before taking the sine.

Interpreting periodic regression output (for phase, amplitude), and obtaining angle from coordinates

Task: find amplitude [kAMPLITUDE(sin,cos)]

```
Function kAMPLITUDE(bsin As Double, bcos As Double)
' amplitude, by Pythagoras's theorem, in the same units as Y. It is
        fundamentally the length of a vector from (0,0) to any (sin,cos)
        coordinate.
    kAMPLITUDE = (bsin * bsin + bcos * bcos) ^ 0.5
End Function
```

Task: Angle or Phase angle, to be retrieved from its sine and cosine [a set of functions]

The function kANGLE(cycle length, bsin, bcos) does several jobs. It returns angle from any (sin,cos), and phase angle from sin and cos regression coefficients. kANGLE also gives the Mean Angle from paired sums (or means) of sines and cosines of unit vectors—but if the data are based on non-unit vectors, or are (sin,cos) of non-unitary Mean or Resultant vectors, then it yields the Resultant Angle. *Cycle length* in this function sets the output as degrees, radians, day of year, etc.

kANGLE is the 'front' function, the usual one used in the spreadsheet. kANGLE calls (farms out work to) the supporting macros. These could have been written as a single macro, but dividing it up makes parts available for other uses, and code divided into parts is easier to check.

```
Function kANGLE(cyclelen As Variant, bsin As Double, bcos As Double) As
      Variant
'USER INSTRUCTION: if 0 entered for cyclelen it is shorthand for RADS
'Returns ANGLE  for specified cycle length, from regression coefficients bsin
      and bcos.  It calls kANGLERADS, then converts to stated cycle length
If cyclelen = 0 Then cyclelen = 2 * WorksheetFunction.Pi()  'shortcut is 0 for
      RADS
If bsin <> 0 Or bcos <> 0 Then 'only calculate if are not both zero
   'CALLS: get angle and convert to same units as cyclelen
   kANGLE = kANGLERADS(bsin, bcos) * cyclelen / (2 *
      WorksheetFunction.Pi())
   kANGLE = kMODU(kANGLE, cyclelen) ' returns result as < 1 cycle
End If
'error message if angle undefined/nonexistent
If bsin = 0 And bcos = 0 Then kANGLE = "#error: sin&cosBothZero!"
End Function
```

kANGLERADS, the core Angle function, takes a sine and cosine and returns the angle in radians. It calls the functions kANGLEONAXIS and kQCrads.

```
Function kANGLERADS(bsin As Double, bcos As Double) As Variant
'Returns ANGLE  in radians from (bsin,bcos) coefficients.  This is the core
      ANGLE  function. Called by kANGLE . Calls Functions
      ANGLEONAXIS, kQCrads. Is of type variant in case an error
      message needs reporting. Input is B1 and B2 (sin&cos coefficients) or
      sin and cos coordinates.
' check TWO special conditions before taking ratio bsin/bcos
' ONE, if both are zero:
If bsin = 0 And bcos = 0 Then 'both are zero, null data submitted
kANGLERADS = "#error: sin & cos both zero!"
'TWO, if one is zero where is the ANGLE :
   ElseIf bsin = 0 Or bcos = 0 Then 'if either is zero, ANGLE  is on an axis
   kANGLERADS = kANGLEONAXIS(bsin, bcos)  ' FUNCTION CALL
   Else
   kANGLERADS = Atn(bsin / bcos) + kQCrads(bsin, bcos) ' arctan + QC,
      FUNCTION CALL
   kANGLERADS = kMODU(kANGLERADS, 2 * WorksheetFunction.Pi()) '
      kMODU so that angles not reported as >= 1 cycle
End If
End Function
```

Handling special case of angle on axis (i.e. equal 0°, 90°, 180° or 270°): because arctan is the key to obtaining an angle from coordinates, two of the on-axis cases can generate division by 0 which is mathematically

undefined. That requires a way to handle such cases. These are usually
called by the function kANGLERADS, not directly by the user.

```
Function kANGLEONAXIS(bsin As Double, bcos As Double)
' Returns the ANGLE in radians if either (NOT both!) bsin or bcos = 0.  This
        function invoked usually by other "ANGLE ..." functions. An error
        message, if BOTH bsin and bcos = 0, should be returned, but if called
        via another Fn the error message may not show).
' first check that at least 1 but not both are zero
If bsin * bcos = 0 And bsin + bcos <> 0 Then
    If bsin = 0 Then 'ANGLE is either at 0deg or 180 deg
        If bcos > 0 Then 'ANGLE is at top, 0deg if condition satisfied
        kANGLEONAXIS = 2 * WorksheetFunction.Pi()
        Else: kANGLEONAXIS = WorksheetFunction.Pi() 'if IF not satisfied
        End If
    End If
    If bcos = 0 Then 'if ANGLE is either to left or right
        If bsin > 0 Then 'if ANGLE is to right
        kANGLEONAXIS = 0.5 * WorksheetFunction.Pi()
        Else: kANGLEONAXIS = 0.75 * 2 * WorksheetFunction.Pi()
            ' if IF not satisfied i.e. bsin<0
        End If
    End If
Else
kANGLEONAXIS = "kANGLEONAXIS is wrong fn: one argument but not both
        must be 0"  'text message reports error
End If
End Function
```

The QUADRANT CORRECTION (QC) is key to obtaining the proper angle
from its (sin,cos) coordinates.

```
Function kQCrads(sin As Double, cos As Double)
'Called by kANGLERADS.  NOTE the special-case handling of on-axis
        ANGLE is not diagnosed here, but in kANGLERADS where some on-
        axis cases cause DIV/0 error.
' kQCrads uses Fisher's conditional equation as implemented in spreadsheet
        formula; the ">=" is a change from Fisher and prevents 0deg being
        converted to 360deg (which is legitimate but not useful). Worksheet
        equivalent:  "=2*pi()*IF(COS<0,0.5,IF(SIN>=0,0,1))"
' work in cycles first, then convert to rads
If cos < 0 Then kQCrads = 0.5
If cos > 0 Then If sin >= 0 Then kQCrads = 0
If cos > 0 And sin < 0 Then kQCrads = 1
'convert from cycles to rads
kQCrads = kQCrads * 2 * WorksheetFunction.Pi()
End Function
```

The modulo function is often useful—as above, e.g. to convert 363.2° to
its equivalent 03.2°—but because Excel's MOD function is unavailable in
VBA, this is our alternative.

```
Function kMODU(number As Variant, divisor)
'#' UTILITY.  Modulo is remainder from a division. Worksheet function MOD
        not available to VBA. The MOD function can be expressed in terms of
        the INT function:
'#' MODulo(n, d) = n - d*INT(n/d)  //whole - divisor*int(quotient)
kMODU = number - divisor * Int(number / divisor)
End Function
```

Task: Finding significance of a cycle (or of added variables)

Determining significance of a cycle (Sec. 6.4.5) in the context of other variables is done definitively using two regressions, one including it and one not. This method is general and can be applied to any group of variables. A "help" function is sometimes useful to write as a macro.

```
Function kuFimprv_help() As String
kuFimprv_help = "'kuFimprv' Gives F for the variable(s) X added to a model.
        Evaluate X(s) using regressions with & without the X(s) of interest.
        p[Fimprv] = Fdist[Fimprv, df_added, dfresidfullmodel]."
End Function
```

Fimprv determines the *F* due to one or more variables; e.g. *F* due to a set of *X*s is determined from one regression with and one without them. Typically the evaluated set is the (sin`*X*,cos`*X*) pair for a cycle. Clear labelling of input variables helps to guide use when using Excel's insert-formula dialog.

```
Function kuFimprv(SSresid_wXs, DFresid_wXs, SSresid_woXs,
        DFresid_woXs) As Double
' Gives an F, Fimprv, for the improvement due to addition of variable(s) Xn to
        a model. Thanks: W.G. Warren.
' You will need SSresid and DF resid from two regressions, one with ('full')
        and one without ('short') the variable(s) of interest.
Dim SSEs, SSEf As Double
Dim DFs, DFf As Single
SSEf = SSresid_wXs
DFf = DFresid_wXs
SSEs = SSresid_woXs
DFs = DFresid_woXs
kuFimprv = (SSEs - SSEf) / (DFs - DFf) / (SSEf / DFf)
End Function
```

As stated in the 'help' macro, *p*[*Fimprv*] carries numerator DF for the number of variables added (evaluated), and denominator DF for the residual of the full model.

Task: Making an equation from regression coefficients [2 functions]

Here is an example that automates a tedious and error-prone task.

Graphing a periodic regression requires an equation, which can be tedious, especially when there are many terms. This automates it. Use the PASTE FUNCTION command to get prompts on the syntax. Select the coefficient values and names, and the macro "kumakeeq" generates a string that is the equation. Copy and paste the result as "values only" to get a text version of the regression equation; you can then copy this and paste it to your word processor.

To convert the equation to a spreadsheet formula, precede it by an "=" and then replace the X label names with the cells for the same variables (your analysis will have required columns for sin`x, etc., so they should be there already) in the same row (see Sec. 11.1.5). This is quickly done. When complete, extend the formula down the range of the cases (usually rows) you wish to use. That creates a column of fitted or predicted values (against actual or arbitrary† x values, respectively) for plotting.

† It is often preferable, for generating a line from an equation, to not calculate fitted Ys for the original Xs if there are very many points, but to paste the new equation alongside 10–20 evenly spaced X values, to obtain predicted Ys that can be copied to a drawing program and traced by a curve without leaving too many points to remove.

```
Function kumakeeq(coeffrange As Range, labelrange As Range) As String
' ©K.N.I. Bell. Makes a formula from coefficients and labels of a regression.
Dim tempstring As String
Dim i, howmany As Integer
howmany = kCOUNTRANGE(labelrange)
' TEXT function syntax in VBA is WorksheetFunction.TEXT(cell,"0.0000000")
tempstring = WorksheetFunction.Text(coeffrange(1), "0.0000000")
' then pick up pairs of +coeff*label
For i = 2 To howmany
tempstring = tempstring + "+"
tempstring = tempstring + WorksheetFunction.Text(coeffrange(i),
        "0.0000000") + "*" + labelrange(i)
Next i
kumakeeq = tempstring
End Function
```

To count cells in a range, we have kCOUNTRANGE (this function is called by kuMAKEEQ()):

```
Function kCOUNTRANGE(data As Range) As Double
' counts cells, neglects blanks, does NOT TOLERATE error messages
Dim i As Variant, count As Integer
For Each i In data
    count = count + 1
Next i
kCOUNTRANGE = count
End Function
```

11 APPENDIX III: STATS REFRESHER IN A RUSH

11.1 INTRODUCTION

What's this section for? Regression is basic to the most useful stats, and to the most useful technique here: periodic regression. This section gives a quick refresher on basics like *p*-values, residuals, and outliers, in the regression context.

This refresher is not intended as comprehensive, but as gentle help on key concepts so nobody has to get stuck when reading this book.

The topics in this refresher can be further researched in almost any basic general statistics text: for your bookshelf try any edition of Zar. For an internet text, Statsoft's electronic textbook is quite readable, and freely accessible[†].

([†]presently at http://www.statsoft.com/textbook/stathome.html).

Don't be surprised when advice varies amongst statisticians. Different is not necessarily contradictory. There are often many approaches, and no statistician knows all statistics. That diversity is positive.

Note: **the symbol α** (alpha) is widely used for two different things that both appear in this book: [i] α, the chosen critical *p*-value for significance in a statistical test; or [ii] α, an angle in the Azimuthal system used in this book (whereas the symbol θ (theta) indicates an angle in the Polar Angle system). The context should show which α is which.

11.1.1. WHAT IS *STATISTICS* FOR?

Rarely asked; but without the answer, why bother at all?

Statistics is an aid to judgement; a safety protocol for using incomplete knowledge. It's useful, because it's rare to have complete knowledge! Statistics limits us to reasonable statements about a *population* when we only have *samples* of that population. In more formal terms, it moderates generalisations that are made to a population from information obtained only from samples.

When don't you need it? If you have measured the entire population, you generally do not need statistics.

Statistics is for bringing rigour, objectivity, and reliability to the process of generalising from incomplete knowledge.

11.1.2. THE VERY BASICS (STATS IN THE WILD WEST)

Statistics is grounded in probabilities, and something (e.g. a difference, a trend) is conventionally declared 'significant' if it represents a sufficiently improbable result under the assumption of randomness or chance.

Everybody does stats, even if they *think* they hate it.

Analogy from the Wild West? A hot dusty day: you swing through the saloon doors, scattering a few flies. A group has been playing poker, but one fellow has had too many aces; at first it looked like luck†, because it is within experience for someone to have a short string of lucky deals, although they began to suspect something was going on††. But it continued and the other players, relying on experience, decide that nobody's *that* lucky†††, meaning luck (chance, randomness) can no longer reasonably explain his winning streak, therefore—they conclude—there is something else going on. The lucky guy has his hands up, and the challengers are checking his sleeve. They might run the lucky guy out of town, or maybe send him out in a box, *because this is the Wild West, they are statisticians, and his luck was a 'significant' departure from expectation based on experience.*

Translation into statistics-ese:

  † that's H_0, the null hypothesis, "nothing going on"
  †† that's H_A, the alternative hypothesis, "something going on"
  ††† based on data and experience, they reject H_0

Thus, *the background experience underpinning statistical inference is observation of stochastic (i.e., subject to chance) phenomena that can be counted and measured.* There is an impressive unity to all stochastic processes, whether they be card deals, coin tosses, dice throws, or natural variables. These processes and their theory provide the background, the null expectations, for statistical inference (Sec. 11.1.7).

11.1.3. OBJECTIVITY

Objectivity is the key to sound inference and decision-making, not only in life but also in statistics.

The opposite is subjectivity: a storm is forecast, but you want to go sailing anyway, so you rationalise that weather forecasting is inexact ... your boat sinks and you drown. You have paid the price for the sin of *"argument to the consequences"*: *pretending it isn't raining doesn't keep us dry.*

Inferential statistics is a process of objective judgement and decision-making; we process results in ways that let us evaluate them against benchmarks, so we can understand (with a little more rigour than in the Wild West) what we see. Key to that is the α (alpha) level (or Type I error rate, see 11.3.6) that we choose as our cutoff for deciding that something non-random is going on in what we observe; it will be discussed further below, but I mention it now because it is key in maintaining objectivity.

11.1.4. STATISTICAL ELEMENTS & THE DISTRIBUTION

Elements in the underpinning of statistics are:

• the *trial, experiment* or *event*: a defined sequence of action and measurement, whether of dice throws or nature

• the *outcome*: what that *trial* gives: its score, as in the outcomes 1,2,3,4,5,6 for a single dice throw, or the outcomes Heads or Tails from a coin toss, etc.

... and I only mention those so I can mention:

• the *distribution*: the pattern of abundance of values of observations or outcomes.

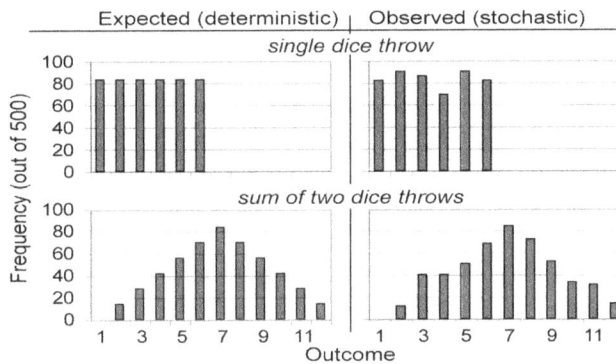

Figure 11-1. Simple distributions. **Upper panels**: a single dice throw. **Lower panels**: a sum of two throws. **Left panels:** expected (theoretically, deterministic); **right panels:** observed (here pseudo-randomly generated). These show two things: [1] Expected and Observed distributions rarely match exactly (so, paradoxically, we expect to not get what we expect); and [2] summing outcomes changes the distribution (summing many uniform variates results in peaked, increasingly normal, distributions). $N = 500$ for each graph. Notice that while outcomes >6 are impossible for a single dice throw, an outcome of 1 is impossible for the sum of two throws.

Distributions are the key to evaluating how likely something is. We can show (Figure 11-1) two very simple distributions to illustrate the idea of uniform (all outcomes have equal probability) and a peaked distribution (central values have higher probability). The sum-of-two-throws distribution also shows the origin of "lucky 7".

The concept "significant" means an improbable[g] departure from expectation. Expectation is obtained from theoretical or real 'null' distributions ('null' reflecting the background known behaviour against which observations are evaluated).

_____

[g] corresponding to an outcome that has low probability of arising by chance

11.1.5. VISUALISING DATA, CASE, INDEX

It helps to visualise a data set as a set of rows and columns. Data represent a sample, or samples, from one or more populations.

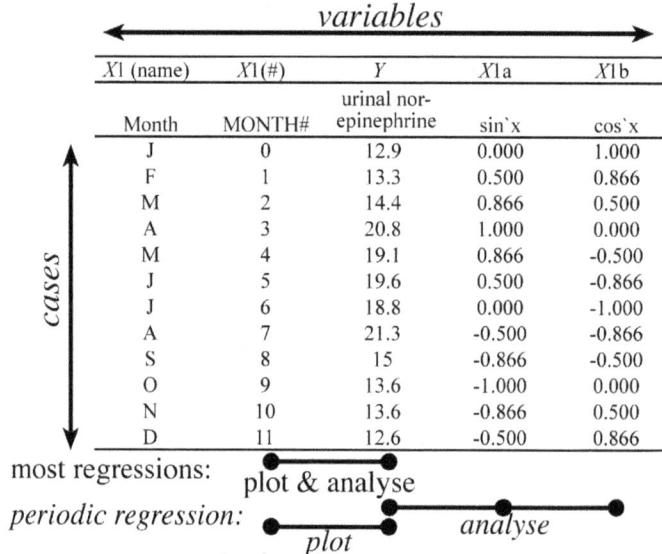

variables

X1 (name)	X1(#)	Y	X1a	X1b
Month	MONTH#	urinal nor-epinephrine	sin`x	cos`x
J	0	12.9	0.000	1.000
F	1	13.3	0.500	0.866
M	2	14.4	0.866	0.500
A	3	20.8	1.000	0.000
M	4	19.1	0.866	-0.500
J	5	19.6	0.500	-0.866
J	6	18.8	0.000	-1.000
A	7	21.3	-0.500	-0.866
S	8	15	-0.866	-0.500
O	9	13.6	-1.000	0.000
N	10	13.6	-0.866	0.500
D	11	12.6	-0.500	0.866

most regressions: plot & analyse

periodic regression: *plot* *analyse*

Figure 11-2. Data as rows (cases) and columns (variables). Plots and analyses are operations on columns. Most regressions analyse and plot the same *Y* and *X*, but *with periodic regression we analyse from transforms of X, and often plot X in cyclic format.* (Data from Batschelet's table 8.2.1, Sec. 9.1; notation `x means properly transformed *x*, Sec. 13.2)

A **datum** (plural is **data**) is one value, one measurement of some kind. One *cell* would contain one **datum**. A **case** usually occupies a *row*, while a **variable** occupies a *column*. Analysis is in terms of variables, sometimes restricted to certain subsets of all the cases.

A **case** comprises one complex observation, e.g. one fish at a particular time (time would be an **index** variable) with **variables** weight, length, nutritional status, activity level, also relating to that time, etc. Arrangements vary according to the question asked.

The number of **cases** (n) in an analysis is what 'gives' degrees of freedom (DF) or, loosely, power to detect effects; the number of **parameters** (m) analysed (count the regression coefficients, except for the intercept) 'uses up' that explanatory power (Sec. 11.3.7), and the remainder can be used for estimating significance, or probability of the result under the null hypothesis.

11.1.6. PAIRED AND UNPAIRED DATA AND TESTS

Regression-like analyses organise data as *pairs* (x_i,y_i) or *cases* ($x1_i$, $x2_i$, $x3_i$, ..., $x[m]_i$, y_i) (Sec. 11.1.5, 11.2.3). A case or pair acknowledges some logical association of the data in that case; e.g. an individual's length, height, and weight; or a set of simultaneous measurements of creatinine, sugar, and electrolytes.

Un-paired data are often used to compare groups or populations but without association of data to other variables at the individual level. E.g. the weights of sparrow eggs in town A, compared with the weights of sparrow eggs in town B). Random sampling is typically assumed. The analysis might use an unpaired *t*-test or an ANOVA (a *t*-test is like a 2-way ANOVA). If, however, you had the daily total weight W of eggs from towns A & B, a paired type of analysis, e.g. regression of $Y = W_A = B_0 + B_1 W_B$, might be reasonable.

11.1.7. PHILOSOPHY OF SAMPLING AND INFERENCE

We sample only to learn (infer) about the population; the sample is of no interest beyond that. You think you see something in sample data, but is it real? ... i.e. does it reflect the population? To answer, ask how likely you'd see it if it wasn't real; i.e. estimate *p* under the null hypothesis.

Inference (Sec. 11.3.5) is all about fairly generalising (making fair inferences) from the sample to the population. Fairness is why statistics constrains inference. Results are assessed, usually, by reference to a (randomly sampled) null population that resembles our sampled population but—crucially—lacks the effect or property being tested for. Null populations are often not explicitly mentioned, but they are implied by the null hypothesis H_0 (Sec. 11.3.6). Indeed, our tables for F, t, Z, etc. represent ideal or theoretical null populations. We assess the probability *p* that random sampling (Sec. 11.3.3) of a null population could produce a result as extreme as our sample; we compare *p* (Sec. 11.2.3) to our chosen α (alpha, Sec. 11.3.1), the significance level for *p* (commonly, we set $\alpha=0.05$). If $p>\alpha$, we consider it plausible as a random sample from a null population (so we do not reject H_0). If $p \leq \alpha$, we consider it improbable in the null population—but should we interpret it as [*a*, fluke] an improbable sample from a null population or [*b*, significant] as signifying a non-null property in the population we sampled? We take a little leap: whenever $p<\alpha$, we favour *b* (significance) over *a* (fluke); so we reject H_0, and allow ourselves to generalise from the sample to the population.

11.2 WHAT IS REGRESSION?

Regression is here, rather than at the end of the refresher, because it's the 'main item'; it will supply context for material further down.

The word "regression" is not intuitively revealing of what it is. Very generally, regression measures association of a Y (dependent) variable with one or more X (independent) variables. It is a way of modelling the variation in one measure by variation in another.

Caution is required, however, as illustrated by that nice old joke about "the price of butter in New Zealand". The joke underscores that [1] because the price of butter like almost anything rises according to inflational creep, it is a trend, and any other trend regressed against it will show some correlation; and [2] correlation may be suggests association, but not sufficient to infer causality. Correlation does not mean that the trend in the height of your apple tree is caused by the butter price. Some philosophy, some theory, about the variables is required to support reasonable inference.

11.2.1. REGRESSION MODEL VS. REGRESSION FIT

A model is a qualitative statement, ideally based in theory, about how Y relates to X. Fitting involves finding parameters by which the model can match the data, and regression is the process of doing that.

We think first of regression as a plot with points (data) and a line (a fitted model) drawn through them. It finds the 'best' coefficients for the variables in some model† that somebody gives to it, that relates Y to one or more Xs, where the Y and the Xs are usually continuous variables.

> †Why do I say "some" model? To underscore that any model is an hypothesis. Furthermore, for any model we consider there is a host of them that we don't. That's a philosophy, intuition, and judgement issue.

The model is a qualitative statement that Y relates to certain terms of X variables (or their transforms). That is the model, built on variables. Intelligent selection of a model form is of course critically important, but substantially intuitive. I.e. selection of a model form is not objective (we have no way for it to be so), and we hope we settle for the next best thing: expert or informed. Our approach is usually via the simplest forms of model. For example, an agricultural researcher might specify a qualitative model $Y=f(X)$ for plant growth by saying it is a sum of functions of three things (the independent variables $X1$, $X2$, $X3$):

growth $- f$(temperature) $\mid f$(soil nutrients) $+ f$(soil moisture)

If the functions are viewed as coefficients (Bs), the model might be written in the general form for a linear model:

$$Y = B_0 + B_1 X1 + B_2 X2 + ... + B_m X_m + \varepsilon,$$

and the same is addressed to any particular point by subscript i:

$$y_i = B_0 + B_1 x1_i + B_2 x2_i + ... + B_m x_{mi} + \varepsilon_i$$

where the B_0 is the intercept, the other Bs are parameters or regression coefficients (the part that is fitted) for the Xs (numbered to m), the subscript i denotes the ith observation, and ε is a normally-distributed error or **residual** (the part, residue, of data left over after applying the model to it; the variation unexplained by the model).

Obviously, if you need only one X ($X1$) then you no longer need to number the (only) X, then all you have is the case subscripts (i):

$$y_i = B_0 + B_1 x_i + \varepsilon_i$$

The regression line itself is described by:

$Y = B_0 + B_1X$

which means the same as the "$Y = mX + b$" (etc.) notation that might be more familiar (but less flexible).

Next comes the fitting part: finding the parameters that enable a match of data with the theory (model). We seek, *for the qualitative model you have chosen to evaluate*, the parameter values that yield the best quantitative fit, i.e. the 'best' parameters B_m. 'Best' depends on a criterion of fit, or 'loss factor' (see 11.1.5); least-squares (Sec. 11.2.2), i.e. minimisation of the sum of squared residuals, is the standard. Each residual, ε_i, added to the regression line would reproduce the original data.

A "best fit" for one qualitative model does not mean your qualitative model is appropriate! It only means you've found a "best" set of parameters for that qualitative model. Your qualitative model should be supportable in theory, i.e. it should make sense; its parameters should be determined empirically, i.e. from data. Occasionally, empirical support appears for a model that doesn't yet make sense, and that empirical support can alert us to previously un-recognised processes. Nevertheless, "empirical support" is not a loophole to support using the activity of your one pet cricket in a matchbox to pick lottery tickets; instead it means a very large number of crickets, chicken entrails, tea-leaves (etc.), before you decide that these auguries that have no theoretical basis might be useful through some as-yet-not-theoretically-understood mechanism.

Generally, regression attempts to relate one *dependent* variable (Y) to one or more *independent* variables (Xs). Think of the Y and the Xs as columns of a table like Figure 11-2, with each row representing one *case* or observation—as in many of the tables you can see in this book. Get used to 'seeing the shape' of the analysis, seeing how variables relate to the columns and how n (number of *cases*) relate to rows.

Regression has great advantages: it is well-known, readily interpreted, and easily conceptualised and visualised. It provides a model or functional relation of Y in terms of X, the model can be graphed, can be probed using residuals, is often useful for de-trending data, and objective tests of significance are available (commonly the ANOVA associated with the regression). It's possible to plot several regressions (lines) on a single graph for comparison.

The simplest situation is one Y vs. one X, both continuous; but there can be many Xs. Other related procedures—to be found in other texts—can accommodate qualitative X variables, e.g.:

ANOVA (analysis of variance), usually (but not always) for the effect of qualitative variables on a continuous Y, or

ANCOVA (Analysis of Covariance), like a regression but having, in addition to continuous X variables, one or more category or qualitative variables, like species E,F,G, ... , etc.

Regression comes in several kinds. Periodic regression (Sec. 6), the main aim of this book, is simply a special kind of regression. Add qualitative X variables and you have a periodic ANCOVA.

11.2.2. LEAST-SQUARES REGRESSION(LSR)

Usually, we just say "regression". Regression has two parts: a model is estimated and then statistically evaluated, typically by (Sec. 11.2.3) the correlation coefficient R^2 and the p-value. The least-squares line (regression function) is usually explicitly calculated, but you could fit it iteratively and evaluate it (or any arbitrary line) statistically by R^2 and p. "Fit" means the fit of model to data, e.g.: we fit (verb) a model to obtain the best explanation of the data; or the fit (noun) would be very bad if the model sloped up to the right while the data sloped up to the left.

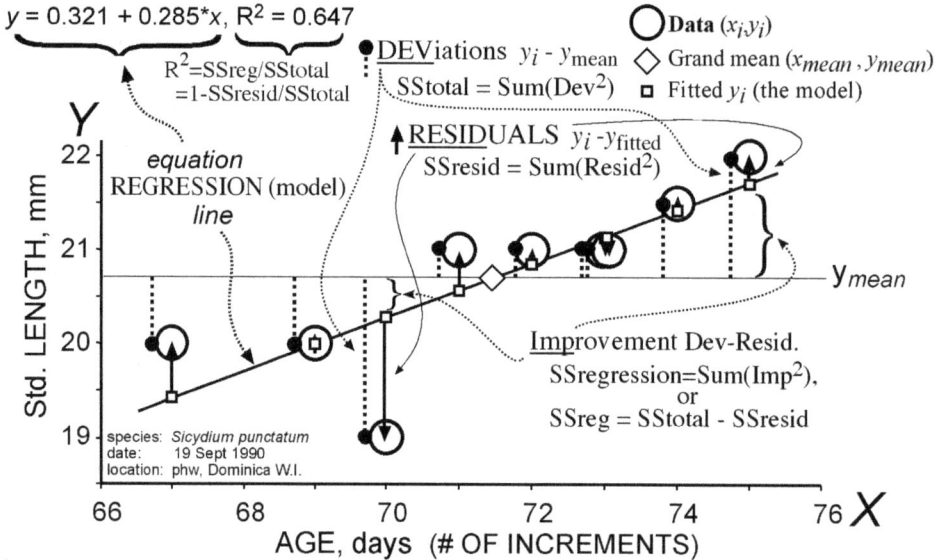

Figure 11-3. Essentials of a regression. For all x_i there are data (y_i), model fitted values $(\hat{y}_i$ or $y_{fitted(i)})$, and residuals $(y_i - \hat{y}_i)$. Length vs. age (days, or otolith increments) for a sample of newly recruited gobies, *Sicydium punctatum*, in Dominica, West Indies, as in Table 11-I. Residuals are always defined in terms of a model being assessed.

We often see the phrase "sum of squares" (SS). There are SS for the total variation (a property of Y alone, no or null model implied), for the regression (a model), and for the residuals (what's left unexplained by the model). The sum of squared residuals ($SS_{residual}$) is a generally useful measure of fit (of a model to the data) and is often used in iterative solving of regressions. Iterative solving involves an adjustment of the parameters and evaluation of the model by the calculation of fitted y values at each x, squaring and summing the residuals to get an $SS_{residual}$, to find the equation that minimises $SS_{residual}$. This minimised $SS_{residual}$ is the Least Squares criterion.

Following Figure 11-3 and Table 11-I: we begin with data, mean y ($\bar{y}$) and the **deviations**, $y_i-\bar{y}$, which we square and sum to obtain SS_{total}. So far, this is without invoking any model. Then, using some regression function—fitted model, $Y=f(X)$, representing a line on which fitted values $\hat{y}_i$ will lie—we calculate **fitted** values $y_{fitted(i)}$ or $\hat{y}_i$ (y-hat-i) as $\hat{y}_i=f(x_i)$; those lead to the **residuals** $y_i-\hat{y}_i$ or $y_i-y_{fitted(i)}$. Squaring and summing residuals gives SS_{resid}. The SS for the regression, SS_{reg} = $SS_{total}-SS_{resid}$, reflects improvement—so SS_{reg} also = $\sum(\text{deviation}-\text{residual})^2$ or $\sum(\hat{y}_i-\bar{y})^2$ —compared to the null model (with zero slope) that asserts $\hat{y}_i=\bar{y}$.

Table 11-I. Raw data for Figure 11-3, and analysis, with notes. **A**: data x and y; mean x and y; deviations (Dev) y_i-y_{mean}; sums of squares total (SS_{total}), residual (SS_{resid}), and regression (SS_{reg}), using regression coefficients in C. Typical regression output: **B**, overall statistical evaluation; **C**, coefficients with significance and 95% confidence limits.

A: Data, grand means, deviations, fitted values, residuals and key SS

X AGE(d)	Y SL (mm)	Dev	Dev^2	Y_{fitted}	Resid	Resid^2	Improv	Imp^2
74	21.5	0.778	0.605	21.42	0.081	0.007	0.697	0.486
72	21	0.278	0.077	20.85	0.151	0.023	0.127	0.016
69	20	-0.722	0.522	19.99	0.006	0.000	-0.729	0.531
70	19	-1.722	2.966	20.28	-1.279	1.635	-0.443	0.197
73	21	0.278	0.077	21.13	-0.134	0.018	0.412	0.170
75	22	1.278	1.633	21.70	0.296	0.087	0.982	0.964
73	21	0.278	0.077	21.13	-0.134	0.018	0.412	0.170
71	21	0.278	0.077	20.57	0.436	0.190	-0.158	0.025
67	20	-0.722	0.522	19.42	0.577	0.332	-1.299	1.687
$\bar{x}$: 71.56	$\bar{y}$: 20.72	SS_{total}:	**6.556**		SS_{resid}:	**2.311**	SS_{reg}:	**4.245**

B: Typical regression output: for significance

			$1-SS_{resid}/SS_{tot}$	SS_{reg}/SS_{tot}
Multiple R	0.80469			
R Square	0.64753		0.64753	0.64753
Adj. R Sq.	0.59718			
SE Est.	0.57454	*Standard Error Of the Estimate = Root(MS_{resid})*		
N, Cases	9			

ANOVA

	DF	SS	MS	F_{calc}	$p[F_{calc}]$
Regression	1	**4.245**	4.2449	12.86	0.0089
Residual	7	**2.311**	0.3301		
Total	8	**6.556**			

C: Typical regression output: for coefficients *(for equation: SL=0.321+0.285*AGE)*

	Coeff.	SE	t Stat	P-value	Lower 95%	Upper 95%
Intercept	0.3213	5.692	0.0564	0.957	-13.14	13.78
AGE	0.2851	0.0795	3.5861	0.009	0.097	0.473

A residual is a difference ($y_{observed}-y_{predicted}$) between a y_i as observed and as predicted ($\hat{y}_i$, y-hat$_i$) by the model (Figure 11-3) at observed values x_i. The ideal fit reduces the residuals to zero, but that virtually never happens†; we carefully choose a model form and then fit the best coefficients for its terms. († A model with too many parameters can consume all available DF, with none left to support significance; a line may have enough bends to pass through all data and give a zero residual SS, but that's like declaring the data to be the model: pointless. A perfect fit indicates either over-fitting, or a deterministic situation that does not need statistics.)

"Least-squares", meaning least sum of squared residuals, is the conventional objective criterion of the best fit for the given model. Thus the regression line giving the least (smallest) sum of squared residuals (Figure 11-3) is considered the best fit to a given model. Least-squares is one kind of criterion of fit, or 'loss factor' as Wilkinson (1987) terms it.

> [Note that although the choice of loss factor is objective in the sense that it is independent of any preconceived conclusion, It is also somewhat arbitrary: some stats programs let you choose, instead of the conventional (Figure 11-3) least-squared-residuals criterion, virtually any other criterion, e.g. the sum of residuals raised to any power. The Systat manual (Wilkinson 1987) explains that its nonlinear estimation routine can use any variable, even a random number, as a loss factor.]

The regression table (Table 11-I) provides the equation ($Y = 0.321 + 0.285*X$), as the intercept plus the product(s) of coefficients (slopes) and their associated X variables. We can recalculate the F as a ratio of the two Mean Squares (MS), and those in turn are the SS divided by their respective DF. We can then look up—in an F table—the p or probability, under the null hypothesis, of an equally extreme F with those DF.

The example regression (Figure 11-3, Table 11-I) is highly significant. How does that work? First, our reference F represents a theoretical null population (Sec. 11.1.7, 11.3.6), i.e. with "nothing [new] going on", where the null hypothesis H_0 is correct. There, the probability of finding a reference $F_{DFreg,DFresid}$ as great as our $F_{calculated}$ is very low ($p[F_{1,7} \geq 12.86]$ = 0.0089); i.e. the probability of an F-ratio that large occurring by chance alone is about one in a hundred. Next, that $p[F]$ is much smaller than the usual critical α of $p=0.05$. Finally, we therefore conclude the result is (highly) *significant of, strongly signifies,* that *something is going on;* there *is* an effect of X on Y. Formally, we reject the null hypothesis. Essentially, by rejecting the null, we've decided (Sec. 11.1.7) it's more plausible that we sampled a population with something going on than that our sample is a highly improbable chance result from a population with "nothing going on". Implausibility of chance as an explanation for a result leads us to conclude that the source population has, in some degree, the property we tested in the sample.

Another way to think of p: it says, effectively, that if you took all the ages and lengths separately in two hats, and sampled by randomly choosing one age datum, then one length datum, then another age, another length, etc. for 9 pairs, and if you did the entire exercise thousands of times, you would get (as $p=0.0089$ implies) an F value as large (12.86) with those DF less than 1% of the time. You would have been using a resampling stats approach, and it would indicate whether the parametric probability is accurate. In terms of the slope, the regression's ANOVA says that in a normally-distributed population (with nothing going on) you would have to randomly sample about a hundred times to get a slope this different from zero. "Luck of the draw" (H_0) is rejected in favour of a relationship between X and Y. The R^2 indicates that 65% of the variation in Y is being explained by X, so the effect is strong and the regression is probably useful if you want to know about the relation of Y to X.

Does it matter which variable is assigned to X and which to Y? Yes! The X or *independent* variable is the one that by logic influences the *dependent* or *response* variable Y, not vice versa. Can we simply invert a regression done backwards to the right way around? No: what is rarely noticed (by us, mere biologists) is that LSR is not symmetrical, i.e. the line for Y vs. X is not just the inverse of the line obtained for X vs. Y. The lack of symmetry is inherent in the definition of the residual as $y_i - \hat{y}_i$ or $y_{observed} - y_{fitted}$ corresponding to the same x (i.e. the vertical distance of an observation from the regression line, Figure 11-3). It is a powerful reason to carefully choose which variable to place on the X axis, and which on the Y. The lack of symmetry can have implications for situations where neither variable is properly either an *independent* or *dependent* variable, for example in length-weight ratios.

Definition of the residual as the vertical distance $y_i - \hat{y}_i$ reflects the usual assumption in regression that the X values have been measured *without error*. That is not always true, but is often in practise ignored. If X is to be treated as incorporating some measurement error, then a symmetrical variant of LSR, the Geometric Mean Regression (GMR), also called functional regression, can be used. In GMR the residuals are the distance from each (x,y) observation to the nearest point on the regression line (i.e. distance to the coordinates of the nearest fitted value). GMR therefore is symmetrical: the residuals given by regressing Y on X are the same as by regressing X on Y. (It's a bit regrettable that the name "least squares regression" isn't a good differentiator against GMR, because although GMR uses a different (x and y, not just y) residual, it still uses the least sum-of-squares of residuals. Really, we mean two groups of regressions: those where residuals are parallel to the Y axis, and those where residuals are perpendicular to the model line.) GMR is not offered as a standard option in any package I know of, but Ricker (1984) explains how to readily estimate the GMR from an LSR.

There exists an infinity of models, or regression lines, that can be imposed on any data set. The form (structure) of the function can be varied, for instance using log-transforms $\ln(Y)$, quadratic (X^2) or higher-order (X^h) terms to find the kind of function that can best describe the data. Remember that LSR does not on its own tell you that the model you have used is of the right type: LSR is merely one (standard) approach to quantitatively optimise a given qualitative model.

Residuals result from both data and a model (line). To arbitrate between competing models you would first look for theoretical reasons to prefer one (i.e. favour simplicity and theoretical basis); then, you could arbitrate on the basis of R^2, p, lack of residual pattern (pattern implies that the model is wrong or incomplete) and normality of residual distribution (to respect the parametric statistics based on normal distributions).

Residuals are a powerful exploratory tool. They can be considered de-trended variables (Sec. 6.4.6, Intercorrelated Xs (e.g. de-trending temperature)). E.g., you might pool data for populations A and B to get one regression and then compare residuals A and B (using t or ANOVA).

Outliers

Outliers are extreme data that are judged to not belong to the population of interest. The motivation to delete them is that their influence on a regression is large, because the least-squares criterion gives more 'leverage' to extreme values. Typically, outliers show poor agreement with the model, but that may say as much about the model as the outlier; sometimes you are better off using transformation (e.g. $\ln(Y+1)$) to reduce the effect of extreme values. Once you remove one outlier and re-plot, you will often see secondary outliers. Keep going and you're left with two data points, a perfect fit (the line joining them), and no significance. The value in identifying an outlier is to give you an opportunity to check for clerical errors or reasons to consider it not a member of the population of interest. Do not exclude "outliers" from a model unless you are ready to explain your theoretical or other logical basis for doing so. One good reason can be that including the outlier led to a less cautious prediction, e.g. one that risked overfishing, etc. Any outliers excluded from a model must be declared in the publication.

11.2.3. WHAT IS p, OR 'SIGNIFICANCE'? WHAT DOES R^2 MEAN?

We want more from a regression analysis than just the parameters of a model. We also want an indication of the regression's reliability and its usefulness. We wouldn't feel safe using it if it wasn't *reliable*; and no matter how reliable it was, we wouldn't bother to use it if it wasn't *useful*.

Reliability is indicated by a low p (p indicates how unlikely a result could arise by chance, and therefore how strongly some determinism, some effect, is *signified* by the data), while R^2 (the squared correlation coefficient) indicates how useful the model is in terms of explaining the variation in Y in terms of X variables.

p: The overall probability p (from which we judge significance) for the regression expresses how likely an equally extreme F would be if you sampled the X and Y independently from a null population, one in which there was "nothing going on" (Sec. 11.1.5, 11.1.7). What's the range of values that p can be? $0 \le p \le 1$, and a p of 0.9 means "90% probability". (There cannot, by definition, be any probability greater than 1.0; $p=0$ refers to something that will never happen, and $p=1.0$ refers to something that definitely will happen, and we generally avoid those in statistics—why?—because they would reflect strong determinism where statistics is un-necessary.) If a p-value indicates that a result was quite improbable, how low would it have to be to indicate, signify, that something is going on between the variables? Some (not all) use a benchmark, a critical value ($p<0.05$ is commonly chosen criterion or α, "alpha", level) to arbitrate significance†, so that a p-value lower than alpha is taken to mean "the result seen is not attributable to chance" (at $p=0.05$, that's a 1 in 20 chance of getting an equally extreme result in data where there is no true relationship).

> † In seminars and papers, the word "significant"—like any word important to statistical conclusions—should be reserved for its statistical meaning, to avoid confusion or any hint of misleading the audience.

Table 11-II. Data from *Sicydium punctatum* in Dominica, W.I., 1989-1991. These data are for young fish recruiting to fresh water and the fishery, and we are looking at two measures of fish length, the fork length (FL) and the standard length (SL).

Choose *Y* and *X*:	*Y*	*X*	*Xshuffled*
case i.d. #	FL	SL	SLshuffled
1	21.7	19.1	18.75
2	23.3	19.75	22.15
3	23.0	19.8	18.0
4	21.4	18.5	17.5
5	22.8	19.45	19.0
...	...	...	...

In a multiple regression a probability is given, as well as for the regression overall, for each coefficient; that allows judging which independent (*X*, predictor) variables are significantly related to *Y*. It is not uncommon for workers to drop variables showing no significance, but there are deep waters around this practise because it holds the danger of finding combinations that are flukes rather than real, and there are complex issues of undeclared penalty DF for fitting these parameters now missing from the declared regression. Those missing DF can be argued to result in an inflated *F* and a falsely greater chance of declaring a result significant. "Stepwise" procedures, which drop or add variables based on their effect on the model, hold special risk of finding pseudo-significance. To quote Wilkinson's Systat manual:

> "stepwise regression programs are the most notorious source of pseudo '*p*-values' ... [those] which print F-tests and *p* values are inviting abuse; statisticians seem to be the only ones who know these are not 'real' *p* values. The stepwise option is provided ... only for the purpose of selecting a subset model for prediction purposes. It should never be used without cross validation." (Wilkinson 1987)

Likewise, if you do 20 analyses using α=0.05 (0.05 = 1/20), don't get too excited when one or two show as significant. You virtually ensured that by running $1/\alpha$ analyses.

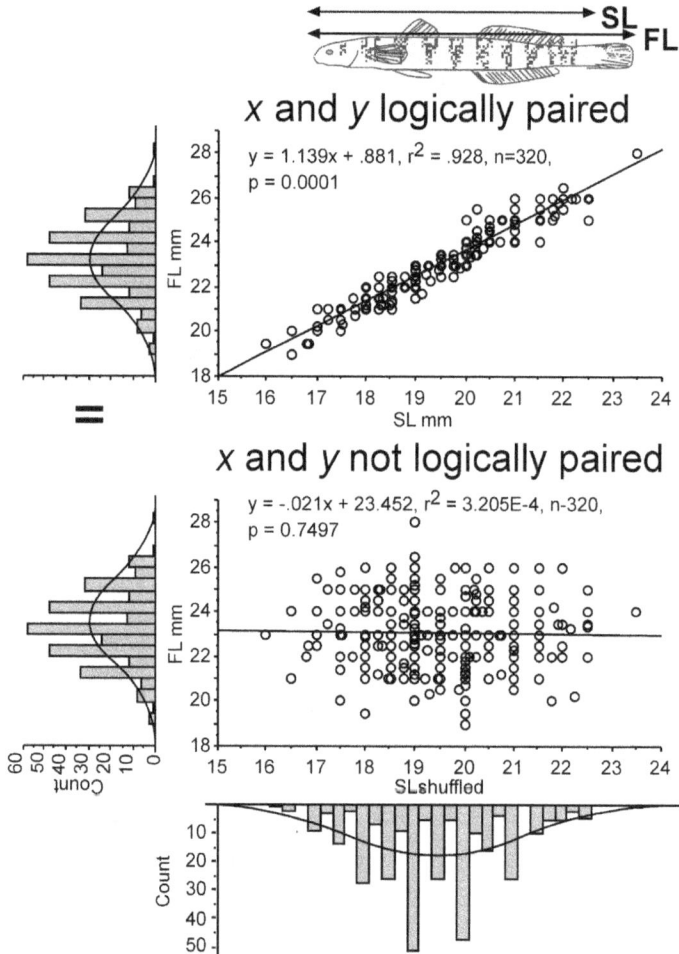

Figure 11-4. The probability (*p*-value) indicates degree of consistency of a result with the null hypothesis. For data in the upper panel (logically paired) *p* indicates, loosely speaking, the probability of obtaining as extreme a result from shuffled or randomly (i.e. not logically) paired data (lower panel). More precisely, *p* (actually *p*[*F*]) is the probability of getting, under the null hypothesis H_0, an *F* as high given the same DF. Frequencies of fork and standard lengths (FL and SL) are given as counts, with curves eye-traced.

***R²**:* A regression's usefulness is indicated by r^2 (often written R^2 for multiple regression), the coefficient of determination ($0 \le R^2 \le 1$), equal to $SS_{regression}/SS_{total}$. R^2 expresses the percent of the [sum of squared] variation in *Y* explained or predicted by variation in *X*. An R^2 of 0.9 is read as "[about] 90% of the variation [in the data] is explained [by the model]". The residuals are the unexplained portion. The Adjusted R-squared, $1-(1-R^2)*(n-1)/(n-m-1)$, anticipates the R^2 that may be expected in further sampling of the same population.

The correlation coefficient r $(-1 \leq r \leq 1)$, is available in bivariate situations. r is a magnitude with sign; only its magnitude can be obtained as $\sqrt{r^2}$; r is therefore calculated $r = \sum xy / \sqrt{(\sum x^2 \sum y^2)}$†. r usefully distinguishes positive correlation (as in Figure 11-4, upper panel) from negative (where positive change in X is associated with negative change in Y). Squared, i.e. r^2, it is the coefficient of determination. Sometimes however, one suspects, r is quoted simply because it is numerically larger than r^2.

> † the summations' reference to lowercase variables conventionally alerts that they are not calculations but abbreviations for the sum of squares of X, of Y, and the sum of cross-products, defined and calculated as:
>
> $$\sum x^2 = \sum (X_i - \bar{X})^2 = \sum X_i^2 - (\sum X_i)^2 / n; \quad \sum y^2 = (Y_i - \bar{Y})^2 = \sum Y_i^2 - (\sum Y_i)^2 / n;$$
> $$\sum xy = \sum (X_i - \bar{X})(Y_i - \bar{Y}) = \sum X_i Y_i - (\sum X_i)(\sum Y_i) / n.$$

Let's illustrate p and r^2 in a regression (Figure 11-4) with real data, and then with X randomised or shuffled (Table 11-II). In the table, SL and FL (x and y) in the same row logically pertain to the same fish; but SLshuffled pertains to any fish, i.e. instead of measuring fish by fish, we put all the FLs and SLs in two hats and picked them at random to destroy their logical association. While we expect substantial correlation of FL and SL logically paired to the same fish, we don't expect much correlation of FL of one fish with SL of another fish. The probability ($p[F]=0.0001$) for the 'logical' regression says that in a null population (like the shuffled one) we would need to randomly draw these n data about 10,000 times to get an equally strong relationship (F). At $\alpha=0.05$ and $p[F]=0.0001$, $p<\alpha$; that F result is incompatible with H_0, so the sample property (correlation, slope of Y vs, X) plausibly exists in the sampled population. Conversely, the probability ($p=0.75$) for the 'shuffled' case says that a slope as strong could often (75% of the time) arise in random data (or a null population), which the shuffled data are, so it is unsurprisingly compatible with its H_0 and cannot signify a relationship of Y to X in the [shuffled] population. The correlation coefficient r^2 is high in the 'logical' case. In the 'shuffled' or null case r^2 is about nil—it's the classic "football" pattern of two uncorrelated variables.

11.3 MORE BASICS

11.3.1. EXPECTATION, OBSERVATION, ALPHA, SIGNIFICANCE

We commonly encounter "expected" and "observed" values. "Expected" values are those implied by known system dynamics. Expectation does not incorporate the property being tested, because it is not part of known dynamics. "Observed" values are those, well, observed; we also call them data. The null hypothesis H_0 is that observations are explainable by random sampling in a null population matching the known dynamics but lacking the property (effect) being tested.

Inference (Sec. 11.1.7) leverages *observation* (data) against *expectation* (under the null, H_0), so that an index of the difference between them can be derived (usually as a probability) and used to determine "significance". Observations that lie close to expectation are "not significant", i.e. do not signify anything but chance interfering; but observations far from expectation are inconsistent or incompatible with H_0, implying or signifying some other effect in the parent population. A slope of *Y* vs. *X* may be apparent in the sample, but the expectation or null assumes the true population slope = zero; only when the sample result is sufficiently improbable under that assumption do we call it significant.

The threshold probability, α (alpha), for "significance" is chosen by the researcher, and must be declared (or inferences become meaningless); it's best however to report not merely "significant" or not, but the exact *p*-values found. Typically $\alpha=0.05$, and a result is significant if $p \leq \alpha$. If you have 100 regressions, or 100 parameters in some number of regressions, you should expect under conditions of "nothing going on" that 1 in 20 would show $p \leq 0.05$, so that would be unremarkable; it would be irresponsible to cherry-pick results for those few grains of wishful thinking.

11.3.2. POPULATION, SAMPLE

My apologies to those who feel talking about sample and population after the distribution is putting the horse after the cart.

Statistically, a **population** is a group of subjects or items of interest, defined by kind, size, geography, etc. To constitute a population, the subjects have to share some key properties or at least behave the same way with respect to the properties we are interested in. We assume the properties of the population can be estimated from samples taken (usually randomly) from the population. Usually we do *not* measure (etc.) the entire population but only samples from it. A **sample** is a group of data from a (presumed, defined) population.

Subtleties arise when you define your population, and when you define your sample. A population can be defined on the basis of any attribute. Definition of the population is often unstated, even though for some study it might be practically limited to, e.g., "the population of snails in my yard", or multiple populations may be implicit from a comparison, e.g. "[population of] snails found on stones" *vs.* "[the population of] snails found on plants", which could make sense for some questions but not for others—i.e. the biology should not be forgotten.

To be valid, a sample should be **unbiased**. 'Unbiased' means that the method of selection is impartial, not expected to result in samples that have consistently greater or lesser values for the variable being studied, e.g. that your sampling of mouse weight does not use a trap that big mice can't fit into, or a trap with an insensitive mechanism that remains un-triggered by smaller mice, etc.

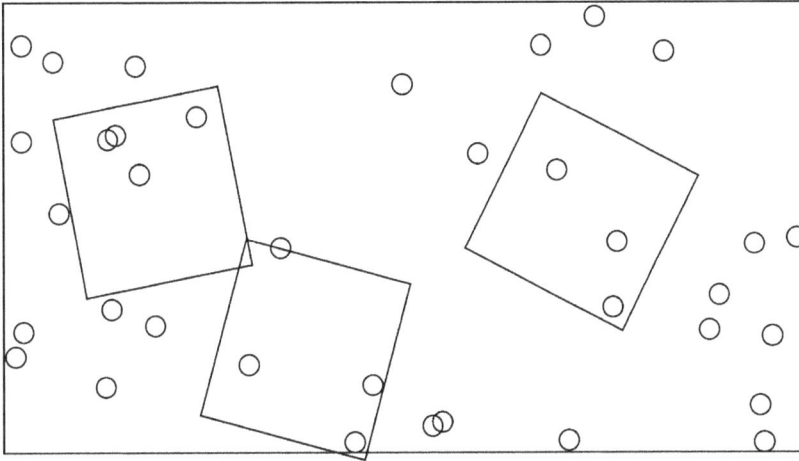

Figure 11-5. A population of pennies randomly distributed on a tray, sampled by randomly thrown square quadrats†. Counts yield only one datum per quadrat, whereas attributes (e.g. weight, year, or dirtiness) per penny would yield multiple data per quadrat, increasing the potential complexity of questions answerable. With quadrat and tray size areas, counts can give a population estimate; 3 such can allow an estimate with confidence intervals††; weights would allow inter-quadrat comparisons and inference.

    [† **Quadrat**: a device of fixed dimension used to frame a sample area. †† ≥3 data are needed to calculate a variance, which is necessary to generalise to 'the' population].

11.3.3. RANDOM SAMPLING AND BIAS

Results can be biased by unanticipated behaviours of the subjects/system under study, or of the experimenter/observer. E.g., for a survey of attitudes, we might pick subjects at random from the phone book—i.e. choose subjects without regard to their feelings about the issues being surveyed—and conduct interviews. So far so good, but if some declined to answer for reasons related to the factors being surveyed, then the sample is biased, no longer random. Extrapolation to the population requires the assumption of an unbiased sample, so, if some subjects declined because of the questions, then some 'safe' attitudes might be over-represented in the sample. Even non-respondents who were not at home when called may share some attributes in a way that biases the sample. *Therefore, no matter what the study, each sample excludes all the fish you couldn't catch, all the people who wouldn't answer their phone, etc., therefore the study's results extrapolate only to the portion of the population you could have sampled: the fish you could have caught, the people who could have answered their phone, and so on.*

The conventional way to obtain unbiased sampling is **random sampling**. Let's say we're sampling *x* (height) in a population. *Random* means the choice is not predictable because all *members of the population* are equally likely to be sampled—that's not the same as saying all *values of X* are equally likely to be sampled, because some values may be more or less common in the population, and it is (we repeat) the members of the population that must be equally likely to be sampled. For any variable *X*, when sampling at random, the likelihood of any given value being sampled must depend on its frequency in the population. Incidence of *x* often follows a distribution such as the normal (a.k.a. Gaussian or bell-shaped) curve, but other distributions are possible.

Random sampling is an intention that's met or approximated by various means, often by assigning random numbers. I.e. instead of physically shuffling all your potential subjects in a hat (difficult if they are, for example, rabbits) and picking them out without looking or getting bitten, you could number your subjects and put the numbers in a hat (a sort of proxy-physical shuffling). Or, forget the physical route altogether, number them and let some other "random" process pick the subjects. "Random" numbers generated by computers are technically "pseudo-random". Lotteries often take the trouble to use mechanical machines to shuffle (*def.*: to re-order out of any predictable sequence) a few numbered tokens like ping-pong balls and generate very long sequences; if you prefer, they randomly sample a small population of numbers—those put in the machine—to generate a long sequence that is essentially unpredictable (and highly improbable, which is why lotteries make money).

To randomly sample by using random numbers, think about their distributions. One dice has 6 (equally probable) possible outcomes (1,2,3,4,5,6), but a double throw (sum of) has 11 outcomes (2,3,4,5,6,7,8,9,10,11,12), with higher probability for central values than extremes (Figure 11-1). The more you sum over many random numbers drawn from "flat" or "uniform" distributions (e.g. 1 to 6 for one dice), the greater the expectation of central values, and the more it resembles the normal distribution.

Thus, random number distributions vary from flat/uniform to peaked, so it's important to know whether you want a uniform random number (as for choosing sample sites on a grid, etc.), or something like a Normal(ised) random number (for simulating processes that favour intermediate values).

The key to random sampling is that the subject/sample is chosen in a way that cannot conceivably be related to the question, to avoid bias in the result. For some field studies, a reasonably sized quadrat tossed carelessly over the shoulder can be adequately random; or a transect (a line along which sampling will be done) can be selected (cautiously, randomly within the stratum or area having the attributes of interest) and quadrats placed along it at distances determined by random numbers. A practical source of (flat/uniform) random numbers is often the last few digits on a column of telephone numbers from the directory; it works because numbers are not assigned alphabetically, so any sequence in assignment has been destroyed.

11.3.4. OPERATOR BIAS PROBLEMS AND "BLIND" REMEDIES

Bias doesn't equate to ill intent. If bias were as blatant as the situation that generated the often used phrase "from the Russian judge", it wouldn't survive the simplest checks in science. Therefore, subtle and inadvertent bias, while it may be more forgivable, is more dangerous because it is more easily missed. This is the same principle that says a mediocre plumber, lawyer, (etc.) is more dangerous than a rankly incompetent one, because the mediocre one takes longer to detect.

Related to the sampling bias problem is the key philosophical† problem of experimental bias. The problem is intrusion of subjectivity into sample choice, results, evaluation, and interpretation. Mere knowledge of the theory of the experiment is often enough to influence the results—no matter how strongly the experimenter wishes otherwise. The influence can come via the subjects (if the subjects in the experiment are sentient) or via the operator (we generally hope that operators are sentient). "Blind" techniques isolate the evaluator from any information that could influence evaluation. The choice is clear: will your experiment measure a real thing objectively with negligible contamination from preconceptions, or will your experiment measure the preconceptions with some contamination by real data?

   (†philosophical concerns are paramount, far beyond simple ones like
   measurement error, because they go to the core of the experiment: the logic
   of its interpretation and the ability to generalise.)

Essentially, in most experiments we are measuring Y and will later (by analysis) relate it to X. What good, then, is any estimate of Y if it is contaminated by X? If we want to measure Y, what good is measuring $Y_{contaminated} = Y + f(X)$? Unless the contamination $f(x)$ is small, relating $Y_{contaminated}$ to X could then be like relating X to X: pointless.

The evaluator should not let prior assumptions influence evaluation of data. Subjectivity—researcher/operator's perceptions—can intrude less easily in small counts (discrete variables of low number), more easily with large counts, or continuous variables, and even more easily with more complex evaluation like reading gels and otoliths. Avenues of subjective bias include sentient operators and sentient subjects (taste testing, ergonomics, or responses of patients). Good science is always alert to the problem and prevents it with some variety of "blind" technique.

Otolith reading (interpreting and counting increments on them to estimate fish age) is a complex process notoriously subject to the problem of subjectivity. An otolith evaluation is supposedly just that: evaluation reflecting the information in an otolith, *not* reflecting preconceptions from fish size, location, year, etc., even the size of the otolith itself. Objective evaluation is difficult enough: increments vary in width and contrast, they are 3-dimensional structures of which the third is apparent only subject to focus (often ignored or misapplied). Otoliths vary in readability, and some just have to be rejected altogether for good objective reasons. Few operators state clear decision rules, and many studies do not employ any blind techniques. A careful researcher however will allocate time to quality, and be troubled by potential background contamination of readings. A simple step is to code all otoliths by number, read them with knowledge of nothing more than that number, and only afterward link up the otolith evaluations with the background information.

Seemingly simple situations can be just as fraught. When counting items within a quadrat, the treatment of organisms at the edge has an unavoidably subjective component ... what is enough to count, a leg, a head, part of an antenna? Of course, we can write a procedure (decision rule) to address edge items, but in practice the procedure adds to operator workload and the opportunity for subjectivity, increasing the need for a countermeasure to subjectivity.

Any complex evaluation needs to be isolated from background data. A clean way to do this is to code the samples so that the operator (reader of the physical sample) is "blind", limiting his access to information beyond the features of the sample itself.

Studies can be subject-side-blind, operator-side blind, or both (so-called double-blind). Subject-side-blind studies avoid giving subjects information that could cause them to act in a way aligned with any hypothesis. Operator-side-blind studies isolate information that has the potential to distort the reader's or operator's evaluation, releasing that information only when the evaluation stage is over.

Experiments that are operator-side blind and also involve non-sentient subjects are effectively double-blind, but conventionally are just called "blind", meaning blind to the (only) sentient side. That usage is justified in terms of preventing the label "blind study" from unduly inflating the credibility of studies on, say, peppermints, where the candies were blindfolded.

The Wikipedia page "Blind_experiment" describes a particle-physics collaboration involving many physicists making difficult evaluations of Y. The problem is that they all have some preconception of the likely result for any given X; they eliminate that influence by evaluating the Y component (the difficult one to evaluate) in isolation from the X, to exclude the opportunity to project their Y-per-X preconceptions onto Y. The X is only brought in afterward. This is like a (too rare) blind otolith analysis where the evaluator sees only an arbitrary ID number.

A diet experiment example: If we have a diet experiment requiring staging of fish larvae and eventually relating growth and survival to a treatment which is given via diet, we are heading for trouble if we as the operator also know what diets are being given. That knowledge itself will fatigue us (if we are seeking objectivity) because of the concentration and discipline required to keep it from mind while working. Suppose our experiment had 20 tanks and 4 treatments, we can isolate the operator from background information by making up all the diets and having a colleague code all the feed bottles (1-20). That would free us from the awareness not only of diet, but even which tanks of fish shared a diet. (Just be nice to the colleague, at least until the results are done ....)

Blind techniques improve quality and reduce operator workload. They rarely apply, however, to the analysis; here, the checks are the statistics together with the declarations, including for example whether and how data were censored or outliers defined. A safeguard (against subjectivity and chicanery) is retention of data, notes, etc., for third-party review if needed, and that's insisted on by many organisations. Scientific detachment, the deep wish to hear what the data actually say, is the key guarantor of quality, the key countermeasure to subjective bias. It isolates not only prior perceptions, but also ambition, from what is published. Too few achieve it.

11.3.5. KINDS OF STATISTICS: DESCRIPTIVE AND INFERENTIAL

Descriptive statistics are calculated means, variances, ranges, etc. that describe a single population. They are the foundations of statistics, but you can't do a lot with them alone. E.g. suppose your samples from population *a* show an average height of 10.2 cm, and from population *b* half a km away show an average height of 11.1 cm; so what? Is population *b* really and truly taller, or could you resample and come up with a contrary result? ... in fact, could you expect, even if there were truly no difference in the populations, exactly the same mean (e.g. 10.658563487cm) in samples? Of course not! ... and that is why descriptive statistics alone can't say whether a difference is real or not. (Actually, no statistics can say that for certain, but only with a given probability.)

Inferential statistics is a set of procedures that asks, in effect, "how probable would this result be by chance alone?" It generates an objective *p*-value: the probability that the difference, trend, etc. in the sample could arise through randomly sampling a null population (Sec. 11.1.7). That allows us to *infer* an effect/difference/trend/etc. (if *p* small, $p<\alpha$), or to *not infer* that there is an effect (if *p* large, $p>\alpha$). We infer an effect by *rejecting* the null hypothesis, and we avoid inferring by *not rejecting* the null. A *significant* result is generally one that rejects the null hypothesis (H_0, under which the results are due to chance).

ANOVAs, *t*–tests, regressions, etc. are inferential statistics because they evaluate whether a difference or trend is *statistically significant*. *"Significant"* is such an important concept in science that we need to avoid the use of the word in its everyday sense if there is any chance that the audience will mistake the everyday word for a claim of a statistical result. A good scientific audience can become quickly—and justifiably—hostile if they think you have misled them (or yourself) by misuse of the word 'significant'.

Inferential statistics objectively asks the question "is there anything going on here besides chance (or randomness)?" Formally, inferential statistics test a null hypothesis (H_0) that says in essence "nothing going on" or "randomness is an adequate explanation for what we see". If what you see could often arise by chance, then the null hypothesis H_0 is *not rejected*. Inferential statistics are intrinsically objective because they rely on a random reference, and objectivity is very desirable because it dramatically reduces the scope for an investigator to 'spin' his results, so that conclusions become independent of social or personal biases. In effect, nobody argues with a highly significant statistical result except if there are grounds to think the procedure was improperly executed, or if data were excluded (or added, or invented ...).

Inferential statistics evaluates an observed result in terms of how likely the observed result could arise randomly; if the result stands in contrast to chance, it implies (lets us *infer*) that something is going on beyond chance. This is an easy concept: nobody buys tickets to watch someone flip some heads and some tails, because that is expected by chance. We all know that it's very rare to flip a coin and get 10 heads in a row ($p = 0.50\string^10 = 0.00097656$), or to roll four sixes in a row ($p = (1/6)\string^4 = 0.000772$), or to draw too many aces .... Even in the wild west (Sec. 11.1.2) they do statistics: the guy with too many aces is shot for cheating (if he really was just lucky, then shooting him involved a Type I error; if he had been cheating and never detected, it would have been a Type II error—but I digress; see Sec. 11.3, 11.3.6).

In essence, inferential statistics boils down observations (data) into a form (result) comparable to other situations similarly boiled down. Inferential statistics calculates results or test statistics from data, then evaluate them by reference to known distributions. For example, regression or ANOVA generate an $F_{calculated}$ and either look it up in an F table to find the probability p (significance) of finding an F as high under the null or random assumption, or we compare $F_{calculated}$ with $F_{reference}$ associated with our chosen α (alpha). We thus can compare calculated probabilities to reference probabilities in known behaviours of appropriate analogues, which may be coins, dice, cards etc., or we can use benchmarks derived from a random population with a particular distribution. Typical test statistics and distributions are the Normal (Gaussian) Z, Student's t, the F distribution, Chi-square, etc.

Recombinations of the data or Monte Carlo simulations (programs that mimic a process) can provide relevant reference probabilities while avoiding the assumptions of known distributions. E.g., if a regression null is "no relation of X to Y", we can randomly recombine: sample X and Y separately to generate a set of artificial pairs and analyse, then repeat this many times to find a probability of a relationship (like a slope) as strong as seen in the real data. Figure 11-4 shows that kind of random shuffling. We could then compare that probability to our old friend α (alpha) and declare the result "significant" or "not significant" (of some effect beyond chance alone). The resampling approach determines the reference criterion (probability of achieving a certain result) from the data; to generalise, it's necessary to assume the data represent the population. By analogy, we can calculate the expectation from a 'fair' coin, but if I want to use a slightly damaged coin I can still determine its behaviour by tossing it many times, and if it tended to fall heads 42% of the time, then any gamble using that coin would have to be based on its own properties, e.g. P(heads)=0.42, and then, using this bent coin, the probability of 10 heads in a row would be $p=0.4200000^{10}=0.0001708$, much smaller than estimated above for a fair (P(heads)=0.5) coin. You will have noticed that the same principle would let us test whether a coin is fair, whether it departs significantly from P(heads)=0.5.

Inferential stats 'raises the ante' for anyone wanting to spin results, and that's a good thing because it helps keep us all honest, and honestly informed. They 'raise the ante' because anybody can present a conclusion without data, or data without analysis, and say "well, *I think* that *my* fertiliser is *way better*" and everybody knows it is merely opinion; but if he uses statistics to answer the question then he must either accept the result or commit a fraud to get the answer he wants. There's little penalty for having a wrong opinion, but big penalties for fraud.

11.3.6. THE NULL, AND TYPE I & II ERRORS

Inferential statistics lets us examine data for effects or relationships between one or more Xs and Y, or for differences amongst Xs, by testing whether randomness (chance) adequately explains observations. Formally the test is phrased as a test of a null hypothesis H_0, though that is often implicit, unwritten.

The *Null Hypothesis*, H_0, is a claim like "nothing going on here, nothing new [beyond chance]", "no difference among groups [beyond chance]", or no correlation of A with B [beyond chance]", or "no effect". H_0 is phrased in the context of the question, which can be quite specific, e.g. "the value of Y is not explainable as a function of variables 1 to 5 [beyond chance]", etc., and these are usually phrased in quantitative terms, e.g. "mean growths due to fertiliser A and B are not different" as "$\mu_A=\mu_B$", or (regression) "all coefficients are zero". The *Alternative Hypothesis*, H_A, is simply that H_0 is not true. The pair of hypotheses thus covers all possibilities, even though the alternative hypothesis may be capable of dissection into many.

> To *not reject* the null hypothesis H_0, or to *accept* it when the result is not significant? Many statisticians cautiously prefer to "not reject" the null. That acknowledges that in many cases the null is "not rejected" simply because the data and precision were insufficient (so far) to give the statistical power to reject it. The wording allows the question to remain

alive. A similar philosophy exists in the two possible acquittal verdicts of the Scottish justice system: "not proven" and "not guilty".

Because we test hypotheses against a random (chance) model, the null really amounts to a hypothesis that "chance *is* an adequate explanation for the observation(s)," whatever they are. The alternative hypothesis H_A always means "chance *is not* an adequate explanation".

Table 11-III. Inference errors occur when the null Hypothesis H_0 is true but called false (Type I error), or false but called true (Type II error). H_0 can be "nothing going on beyond chance", "player not cheating", etc. In the wild west card-cheat analogy (Sec. 11.1.2), *not shooting the lucky player*, and *shooting the cheat*, are not errors; but shooting the lucky player is a Type I error, and not shooting the cheat is a Type II error.

	H_0 actually TRUE *(nothing going on, no effect)*	H_0 actually FALSE *(something going on, effect present)*
H_0 called true *(do not reject)*	correct	error **Type II** *(e.g. false negative, fail to detect effect, let cheat escape)*
H_0 called false *(reject)*	error **Type I** *(e.g. false positive, illusory effect, arrest an innocent)*	correct

Any null is either true or false. We hope to determine that correctly by following objective criteria that lead to *not rejecting H_0* or *rejecting* it. If we get it wrong, our conclusions are one of two kinds of error. To *reject* the null hypothesis H_0 when it is *true* is to declare an effect or difference where there is none; we call that a **Type I error**. Conversely, to *not reject* the null when the null is *not true* is to fail to detect an effect; we call that a **Type II error.** Type I and II errors correspond to false positives and false negatives, also known as α errors and β errors. We defend against Type I error by using a low α (alpha, the type I error rate, the probability value above which the null is not rejected). We defend against Type II error by ensuring that the power of the test (β) to detect an effect is adequately high (key steps are n and precision; the fewer the data, and the sloppier the measurements, the larger an effect must be in order to be detectable). To remember what type I and II errors are, think "null hypothesis is true or false, get that wrong and make errors type I or II".

A null hypothesis is typically evaluated by a *p*-value calculated from sample data and reference tables (which we can think of as representing null populations). The *p*-value is the probability of obtaining as extreme a statistical result by chance (i.e. by random sampling) from a [null] population conforming to the null hypothesis. When *p* is smaller than α we call the result statistically significant, meaning it signifies (points to) inconsistency with the null, and we therefore reject H_0. Bear in mind, though, that the commonly used value of 0.05 for α means that in 20 tests on random data we will tend to find one significant result, which means that result is nothing to write home about. It is good practise to report the calculated *p*-values of tests rather than merely declare them significant or not. Significant calculated values of F, Z, t, etc. can be flagged by asterisks according to a key given with the table.

We minimise our chance of committing a Type I error (false positive, saying something is going on when it isn't) by choosing a low α (alpha) value, the Type I error rate or critical p-value. The test p-value must be lower than α in order to objectively reject the null. Typically, chosen α = 0.05. Setting α=0.05 means 5% of the time is the limit of our willingness to reject a true H_0 (to commit a Type I error); a stricter level like 0.01 is often employed.

Thus, for a regression, an H_0 could say "A and B (populations or variables) are not correlated", and a regression yielding p=0.482 would lead us to *not reject* H_0. In other words, the resulting p=0.482 says, approximately, "you could get a result this strong 48.2% of the time by shuffling X against Y in these data" (see Figure 11-4). At the usual α = 0.05, a result p=0.482 is not significant, therefore do not reject H_0. The sequence of steps generally is like:

1. declare H_0, the null hypothesis

2. choose critical level α (otherwise judge result directly according to p (Warren 1986))

3. analyse sample data (inferential analysis like regression, Anova, etc.) and calculate test statistic, e.g. F_{calc}

4. obtain p of finding a test statistic (e.g. $p(F)$, $p(t)$, etc.) as extreme as F_{calc} under H_0, i.e. in a 'null' population (the tables for F, Z, t, etc. represent ideal null populations)

5. A: reject H_0 if $p<\alpha$. (If our result is sufficiently improbable under H_0, we say it's more probable that we have a non-null population (Sec. 11.1.7)).

    B: If $p>\alpha$, do not reject H_0. (If we could easily get our result under H_0, so our data are consistent with H_0.)

There are two ways to minimise the chance of a Type II error (failing to detect something going on). Firstly, we could raise α (alpha), but that would also raise the Type I error rate. Secondly and preferably, we minimise Type II errors without increasing Type I error rate by ensuring that we have an adequate sample size (that relates to the issue of statistical power to detect an effect), and by minimising measurement (etc.) errors. This should point you to a tactic sometimes used by people who would "rather not know": do sloppy work, have small sample sizes.

11.3.7. DEGREES OF FREEDOM (DF) (NOT A POLITICAL STATEMENT), AND 1- OR 2-TAILED

I place these two topics together because you will often see things like:

A.   "$F_{df1,df2}$ = ... = 12.86", or "F = 12.86 (DF = 1, 7", etc., referring to an F_{calc} i.e. an F value calculated from data, and:

B.   "$F_{a(1),df1,df2} = F_{0.05(1),1,7}$ = 5.59", referring to F_{ref}, the *F statistic* as looked up in a table. It means "the reference *F* corresponding to a (one-tailed) critical probability of 0.05 [a typical α (alpha) value or critical *p* value] with 1 [numerator or DF$_{regression}$] and 7 [denominator or DF$_{residual}$] degrees of freedom [DF] is 5.59".

Part A, F_{calc}, is what your data say. Part B, F_{ref}, is the 'value to beat' in order to declare a result *statistically significant*. (A higher *F* beats a lower *F* having the same DF, but a lower *p* beats a higher *p*.) Part A has no statement of probability, while Part B states the α (alpha) level chosen. The DF must match in both parts.

Part A is then compared with Part B. If F_{calc} is greater than F_{ref}, the reference *F* associated with the critical alpha value (Sec. 11.2.3, 11.3.1), then you call it statistically significant. If not, you call it "*n.s.*" for "*not significant*" and you '*do not reject*' the null hypothesis H_0, which always has the meaning 'no difference,' 'nothing going on,' etc.

In this example (Table 11-I), $F_{calc} > F_{ref}$, 12.86>5.59, so we reject H_0. We should (e.g. Warren 1986) not merely say "significant" but report *p(F)*, the exact *probability* of meeting or exceeding your F_{calc} if the null were really true. Stats programs typically provide this exact *p* (0.0089 in Table 11-I). Then a concise report gives F_{calc}, DF, and *p* in a format like "was/not significant (*F*=...; DF=1,2; *p*=...)".

Similar probability statements appear for *t* and other statistics; not all have the same complexity in DF, but the A-vs.-B comparison, or expression as *p*, is typical for parametric statistics.

Degrees of freedom (DF) = "*says who?*"

Degrees of freedom, commonly abbreviated DF, *df*, d.f., etc., are the statistical equivalent of "*says who?*" DF therefore are very important.

The idea of DF is that if we weigh one potato in a bag, that's not enough data to let us generalise at all about the other potatoes in the bag. If we weigh one bag, we can't generalise to other bags.

DF account for the number of data (**n**) you have, against (minus) the number of things, parameters, or predictors (commonly denoted **m**) you estimated; what's left (typically *n-m*-1) is available to assess probability. DF can be confusing; mistakes can occur in calculation (meaning simple errors including forgetting the penalty value of fitting a parameter before estimating *F* and significance, see under Sec. 6.4.1), or in philosophy where the independent variables submitted to the analysis are incorrectly assumed (as in pseudoreplication, Sec. 11.4) to include all *X*s that plausibly affect *Y*.

Thus a regression has *n* data and fits *m* parameters, and those give the DFs. Using for example the data in Table 11-I: we have 9 cases, each of which is one fish's age (*X* in days) and standard length (*Y* in mm), in a regression. The regression is *Y* = B$_0$ + B$_1$*age; *n* = 9 (just count the data); *m* = 1 (only one slope coefficient, that for age, being fitted); the *total* DF is

$n-1=8$; *regression* DF = m = 1; *residual DF* equals $n-m-1=9-1-1=7$ or $DF_{total}-DF_{regression}=8-1=7$. "Regression DF" does not include the intercept, but one more DF is always taken away (penalty) from residual DF just the same.

What use are DF? DF help prevent you going too far out on a limb, making unsupportable inferences from insufficient data.

When do we apply DF? DF are not needed in the fitting of the function, only in estimating significance.

The usual overall test statistic in regression or Anova is F. Follow along with any regression output table, here Table 11-I. F is calculated (F_{calc}) as a ratio of two Mean Squares: $MS_{regression}/MS_{residual}$[†], then looked up—in Excel it's "$=FDIST(F_{calc}, DF_{regression}, DF_{residual})$"—to find $p[F]$, the probability ($p=0.0089$) that an F that large could turn up in a comparable nothing-really-going-on situation. Or, in a conventional F table, locate (as in Zar) first the page with the appropriate $DF_{regression}$, then find the $DF_{residual}$ down the left column, then read across to find the nearest but not greater table F value (12.2, $F_{table} \le F_{calc}$); then, finally, scoot back up to the "answer" row for the corresponding significance or p-value (1-tailed $p \le 0.01$).   The table lookup method is a little more conservative, $p[F_{calc}] \le p[F_{table}]$, which is why we wrote "$p \le 0.01$", unless we interpolate.

† The ratio of Mean Squares, $MS_{regression}/MS_{residual}$, has a numerator and a denominator, hence the terms numerator and denominator DF.

Stats is filled with reference values and test statistics like p, F, Z, t, r; some are reported with DF, some aren't. It is easy to get a little bit of mental whiplash. DF are needed for calculations or test statistics that are to be looked up (by table) to convert to probability (p) in a null situation. If p(that result in a null situation), or $p(F$ in the null), etc., is very low (lower than our α), then (Sec. 11.1.7) we find it not consistent with the null, therefore we reject the null. Lower p-values are produced by *higher* values of F, Z, t, r, and many other test statistics that (often) are steps along the way to finding the calculated p-value. (Analogy: the further [higher F] you walk along a limb, the less likely [lower p] it will support you.) Degrees of freedom apply to many of these intermediate metrics, but a p-value relates to one analysis after the DF have been taken into account.

In reporting (e.g. in a publication or seminar) we usually give a very concise summary of the values underlying the p-value: F and the DF, e.g. for norepinephrine in Sec. 9: "the relationship was highly significant ($F_{(2,9)}=18.68$[†], $p=0.0006$***)", where the *** is an optional flag for very high significance, reflecting low probability of getting an F statistic that high (≥ 18.68): $p=0.0006$ or 6 times out of 10,000 random shufflings of the data. The p-value, far lower than α, is 'significant', meaning 'this result is very unlikely to result from chance alone', or the low p signifies inconsistency with the null hypothesis H_0 (chance), so we reject it.

† In that example, norepinephrine in Sec. 9, you can read "$F_{(2,9)}$=18.68" as: "F calculated from the data††, with regression DF = 2 and residual DF = 9, is 18.68, with p=0.0006"; the n is probably not necessary to report but it's good to note that residual DF = n − $DF_{regression}$ − 1; then the p-value associated with that F_{calc}, with $F_{(2,9)}$=18.68, is p=0.0006. [†† follow calculation of F in the regressions of Sec. 9 or Table 11-I.)

What are "1-tailed" AND "2-tailed" tests?

The way "hypothesis testing" works is that a complex test statistic like a t or F is calculated from the data. Then, that F_{calc} (etc.) is looked up in a table (or via a stats program, or spreadsheet like Excel, etc.) with the purpose of finding the corresponding probability that the value found could arise in a null population where there was truly no correlation, difference, etc. (i.e. where nothing was going on).

Then a decision has to be made: should this be a one-tailed or two-tailed test? A given F or t statistic relates to either a one-tailed or a two-tailed probability, e.g. $F_{1,11}$=6.72 relates to either $p_{1-tailed}$=0.025 or $p_{2-tailed}$=0.050. Why? The test is based on something you can visualise as a bell-shaped curve (central values with high probability, tail values with low probability), you have to ask whether (given your hypothesis) the difference etc. could reasonably be in the negative tail as well as the positive tail. This isn't always easy, but that at least is the basis of it. If your question is "is A>B?", your null is "A is not greater than B", and the alternative is A>B; that would be a 1-tailed test. But if your question were "is A=B?", your null is "A=B" or "A−B=0", and the alternative is "A≠B" or "either A>B or B>A"; that would be a 2-tailed test. In regressions, the overall significance is usually a one-sided F-test, whereas testing for the significance of each parameter is a 2-tailed t-test. (Excel's function FDIST(F_{calc}, $DF_{regression}$, $DF_{residual}$) returns 1-tailed p; function TDIST(t_{calc}, $DF_{residual}$, [tails: 1 or 2]) requires the user to specify 1- or 2-tailed probability for the given t_{calc}.)

11.4 SUBTLETIES, TRAPS, AND MISCONCEPTIONS

A few key issues generate confusion, errors, and unease with statistics.

bivariate plots: great for exploring simple relationships, or projecting modelled Y against one variable (while holding certain other values constant). But a multi-variable relationship generally cannot be explored or analysed with bivariate plots, so for those we have to read regression tables. Even 3-D plots are often insufficient. Just because a Y shows no trend when plotted against one variable *does not mean there is no effect* if the system is more complex.

small samples and Type II error: Type II error occurs when H_0 is false but is not rejected; in other words when something (beyond chance) really is going on but your analysis fails to detect it. Small samples have low statistical power to detect effects, and it is legend (not myth!) that small-sample studies are a clever way to generate negative, nothing-going-on, conclusions that look scientific but do not do the job of detecting effects, e.g. of deleterious effects when somebody really doesn't want to know.

forgotten variables: some people do their sampling without recording time of day. It costs nothing to record time of day, and it may well explain some component of variation due to counted items being less or more visible, or more readily sensing the experimenter and avoiding being seen or caught. Likewise, I have known people insist that it is a waste of time to record the condition of animals in a feeding experiment. Multi-dimensional data have superior durability because they permit deeper interrogation than merely bivariate data.

intercorrelated independent variables: this relates to operation of an analysis; confounding variables is a similar concept. The problem of intercorrelated variables in the model is often overlooked, but can cause misattribution of effects to other variables. The questions for intercorrelation (every variable has some chance correlation with every other variable) are "how much is too much?", and (if it is too much) "can one be de-trended against the other?" to yield a new independent variable that will be anomalies of some kind.   For example, to include temperature in a model that also included the seasonal cycle, a seasonally de-trended temperature (i.e. the residuals of a temperature vs. season periodic regression, like Eq. 6-1) would be needed because temperature is itself substantially determined by season.

confounding variation: use of this term varies, and some popular texts do not mention it.   It is similar in concept to (*see*) intercorrelated X variables, except that it relates to correlation of some X variable not in the model with one in the model (some use the word in the sense of both Xs being in the model, but that is covered by intercorrelated Xs). One X (or potential X) influences another X in the model; both can plausibly directly affect Y. This is the "price of butter in New Zealand" type of problem; an amusing example was a tongue-in-cheek analysis of the (declining) birth rate in Europe vs. (declining) number of storks (Höfer et al 2004)—both make the point that time is the underlying variable that should have been in the model, and apparent success of a model based on intervening variables is virtually guaranteed, yet it is also guaranteed to be misleading.   The consequence is that the model is qualitatively wrong because it lacks a key variable.

spurious correlation: that which is meaningless because it cannot but arise, because of the arithmetic. E.g., imagine regressing results A (results in category A) vs. $B=1-A$ (results not in category A). Of course, A and B will be negatively correlated! Such regressions are meaningless. Some degree of spuriousness can arise where the same variable is involved in both x and y calculations; but judgement is required.   I.e. "spurious correlation" is not a club to be used uncritically: Ricker (1975) points out "It has frequently been pointed out that a regression of the type Y/X against X is statistically suspect. With X appearing in the denominator of the first variable and in the numerator of the second, random variability will tend to generate a negative slope (curved, it is true) in the absence of any real relationship. However, when a relationship of some consequence does exist, the random component adds little to the slope of any *straight* line that is fitted."

uncorrelated does not mean "unrelated". The statistical word "correlation" means a linear relationship, a degree of one-to-one predictability of Y by X. The absence of correlation in this sense however does not mean there is no clear link between Y and X, only that there's no *linear* relation! For example, the sine and cosine are generally said to be uncorrelated with each other, but as both are functions of some x, it is clear that they are related by those functions. It's just that the functions are not linear. Want to try it? Take some random numbers (in the range about 0 to 2π), take sine and cosine, and attempt to correlate them, i.e. plot cos vs. sin; when you stare at the circle that results, you'll get it.

pseudoreplication (Hurlbert 1984): this nice word confuses a lot of people. Sometimes, indeed, it's used as a blunt instrument by people who barely know what they're talking about, but it's an important issue. It essentially refers to mis-attribution of variation, mis-assignment of DF (degrees of freedom) from one source of variation (e.g. individual error) to a treatment for which, often, no adequate DF exist. Suppose someone tries to compare growth on diets A and B using two tanks of 20 fish each (diet A in tank A, diet B in tank B), tells the stats program to compare ($Y=$) growth on ($X=$) diets A and B. The stats program will spit out a result, but it does not know each group was in a separate tank, it assumes no potentially important factor other than diet that differed between the groups. I.e., it organised all difference in Y around the difference in the only X named (diet), because it wasn't told about the two tanks (it assumed that all fish lived in the same tank and obeyed instructions to consume diet A or B). There may indeed be a significant difference between the two groups, but it relates to both *tank* AND *diet* together (if that were declared as the analysis structure it would be non-orthogonal), with no way to disentangle them. Reviewers will dismiss the result, no matter how nice the p-value looks, and say simply that it is pseudoreplication. That will usually be correct but, when talking to someone who doesn't understand what 'pseudoreplication' is and doesn't really want to know, the discussion may not be constructive. It's not that each individual fish is not independent within the treatment, but that there is more than one common dependency, both diet and tank, amongst fish in A or B. A more convincing challenge to the two-tank experiment is "how will you attribute the difference to diet, and not simply to the tank, or its location, or its water quality, or some other 'tank' effect"? ... the problem being that there are no DF to support the appropriate term to separate them. The experimenters' only option is to declare it as an assumption. The assumption may be defensible on some general grounds in some cases, but cannot be statistically evaluated because the tank effects cannot be evaluated: they have no replication, no DF to use as a basis for evaluating the tank effects. (Therefore, if the assumption was reasonable, it should have been declared in the paper to avoid provoking reviewers' proper objections.) Whereas, with 10 tanks of 4 fish each there'd be the same total number of fish but in each of two diets there'd be 5 tanks†; it is possible to statistically assess the chances of all tanks erring in the same direction by a certain amount or more, and that can allow the experimenter to convince readers that the conclusions are reasonable. (†Of course, the treatments should usually be spatially interspersed, for instance by using a Latin Square design, otherwise you might simply be replacing a "tank" effect with some other effect like "room" or "nearness to the window" etc.)

As an analogy: we make backups of computer files to protect against loss through various hazards. If we make 3 copies of a document on one device, we are protecting (hedging) only against one file written on a part of that device that might oxidise, de-magnetise, etc., but a loss affecting that device will affect all 3 copies. To protect against device loss (and the lower-level within-device hazards), we could replicate at the device level, and have >2 devices, but if these are at the same location a fire could affect them all. Replication in buildings a block away would then give some protection against local fire (and all lower-level hazards), but not against, say, a flood that affected the entire neighbourhood. And so on. Likewise, our replication should reflect (give DF for) the hazard (effect, or source of variation) anticipated. Pseudoreplication is then like having 3 copies of a file on one device and expecting that to protect against a device loss or a local fire or higher-level hazards.

Latin Square (a form of interspersion, for example see around p. 194-5 in Hurlbert 1984): an assignment of treatments so as to reduce the influence of a non-uniform environment (in which the experiment has to be carried out) on the results. The Latin Square or interspersion approach can be applied at the sample or treatment levels, as an alternative to or even in combination with randomisation that could—if used alone—sometimes lead to unfortunate combinations that bias results. In fact one could say that randomisation *must occasionally* generate unfortunate spatial (etc.) biases, so, when using randomisation, it is worth scrutinising the random plan before the experiment and do a new randomisation if necessary.

regular sampling: regular sampling leads to more balanced statistical models and is desirable from that point of view; but (Sec. 6.4.6) it has a downside: it may be logistically impossible if multiple timescales are to be addressed, or in long studies. Non-regular sampling accommodates logistical difficulties, and if as-convenient sampling results in near-random distribution in time, it is unlikely to generate artificial frequencies or interfere with natural ones. Lack of balance can be to some degree mitigated by increased number of data.

averages in analysis: averages are often used by people who don't realise how much they are losing, compared with using the full original data. A regression based on averages may well have a higher R^2 than one based on individual data, but the *p*-value might be lower (because *n* and the residual DF then come from the number of averages, not the original number of replicates). Instead, use (or record) data at the finest scale unless there's a clear reason not to (if it can't be clearly stated, it probably is not a good enough reason). In a similar vein, as a student I once pondered whether to use individual animals in a feeding experiment, or to use multiple animals to reduce the variance; Ian McLaren after a few seconds' thought agreed individuals would generate more variance, but "it would be *good variance*". And so it was.

binning of data along a continuous variable: a sin that's a cousin of "averages in analysis". If you don't have a really good reason to bin data, sufficient to counter the problems it causes, then you should not bin the data. Consider binning a range, say, from 0 to 1, say $0 \leq x < 0.25$, $0.25 \leq x < 0.5$, $0.5 \leq x < 0.75$, $0.75 \leq x \leq 1.0$. It's usually a bad idea. It generates error in the data, because it implies 0.25=0.49, and that (0.25-0.249)>(0.49-0.25) or 0.001>0.24, which is clearly wrong. The true errors (residuals) will be mis-estimated. Such imprecision reduces the ability to detect an effect. In regression, there is no reason to bin data. Of course, to bin circular data is doubly foolish because 1.0 cycles (or 360°, or 2π radians) is the same angle as 0 and the zero is arbitrary; only in trigonometric format (sines and the cosines) is there a linear behaviour. Likewise, some poor studies claim (but fail) to analyse seasonality by binning data (e.g.) to four 'seasons' and ignoring the circular essence of the data.

categorisation of continuous variables: this is a step worse than 'binning'. It reduces a continuous variable to a category set like "low, medium, high". Such categorical variables won't be seen as ordinal data (Sec. 3.4) by the ANOVA, so they might as well be "aspirin, tylenol, dirt".

poor archiving: too many people waste data. They publish based on interesting data with unexplored dimensions, meaning the data have been only partly used/expressed. They then lose track of the data (true examples: 'oh, it was on a computer that broke'; 'contact my student and see if he has it'; 'I think it was in a box that got lost when I moved a few years ago'; etc.). Every institution needs an archiving policy or facility (some granting agencies now require it); it can accept data as open or confidential subject to the researcher's permission.

poor documentation: as with the topics "averages in analysis" and "binning of data", too few papers present data in a form that is raw enough to let the reader re-analyse it or scrutinise the result. The reader then cannot trust the result, because nothing is available but conclusions, unconvincing assertions of significance, etc (it always is an assertion, but made credible only by proper disclosure of methods and (about) data). If the paper doesn't have the depth to support trust of the reader, it will be less used and cited.

normalising data: only the data on which a statistic is calculated needs to conform to the assumed distribution. For example, in parametric regression it is not necessary that the data themselves be normal, but only that the residuals be normally distributed (or whatever error structure is assumed by the test). Only if the residuals failed to be acceptably normal would there be any reason to normalise or transform data—usually, Y; e.g. by taking $\ln(y)$, sqrt(y), y^n, etc., to try to achieve normality in the residuals.

transformations: many students initially abhor transformation. They usually come to realise what its place is. If they instead disdain statistics because of misapprehension that transformation is 'cooking', that disenchantment is tragic in the truest sense because it arises out of a respect for verity and for the data, a respect that is essential in a reliable data analyst. Transformation (e.g. logarithmic) is often necessary to bring data towards a distribution that allows analysis; it doesn't force a result, it merely makes the analysis legitimate and meaningful. Transformation of a circular variable into its sine and cosine is the only way to analyse it; indeed it can be argued that the sine and cosine are the fundamental forms and common or angular format are the transforms.

often more than one right answer: it can be frustrating that advice often varies amongst sources. But there are many ways to skin a cat, variation need not imply disagreement, and the diversity of approaches is positive; understand the spectrum of advice and philosophy, and make an informed choice.

11.5 QUALITY CONTROL ON ANALYSES

The researcher's judgement (Sec. 8) is essential in determining that an analysis was reasonable, proper, and reliable. A substantial portion of that is simply recognising what generalisations are and are not appropriate from the analysis.

For regression, the key things to check for (often by inspection) are (a) remaining trend in residuals in relation to original X variables, and (b) normality of residuals. Inspection is aided by plotting for trends, and by the normal probability plot (Wilkinson 1987). These help to indicate whether the model is reasonable, whether a (different) transformation is required, etc. Residual plots ($[y-\hat{y}]$ vs. x_n) and normality checks become much more important when analysing complex relationships. To moderate worry, here is an experienced quote on quality control ('criticism'):

"For the unwary, there is an inherent danger that is caused by the recent explosion of available methods for criticism. If every recommended diagnostic is calculated for a single problem the resulting 'hodgepodge' of numbers and graphs may be more of a hindrance than a help and will undoubtedly take much time to comprehend. Life is short and we cannot spend an entire career on the analysis of a single set of data. The cautious analyst will select a few diagnostics for application in every problem and will make an additional parsimonious selection from the remaining diagnostics that correspond to the most probable or important potential failings in the problem at hand.

"It is always possible, of course, that this procedure will overlook some problems that otherwise could be detected and that the urge to always apply 'just one more' diagnostic will be overwhelming. The truth is: If everything that can go wrong does go wrong, the situation is surely hopeless." (Cook & Weisberg 1982)

Cross-validation is powerful: apply the analysis either to similar data that have not yet been analysed and compare the results; or subset the data (which can be done multiple times) and re-run the analysis, to see how consistent the results are, e.g. Sec. 6.1, where a subsample of 17 data matches well the results from over 9,000 data. Don't judge an analysis by agreement of its conclusions with the literature, because the literature is not comparable in that way; each study (including yours) is supposed to be independent, which it couldn't be if conditioned by other studies.

Monte Carlo simulation, or analysing randomly generated data, allows probing for structural features in the calculations that lead to an "effect"; if so, some portion of an observed effect may be called spurious.

Special cautions for periodic regression, i.e. avoiding some mistakes easily made, are given under that section (Sec. 6, 6.4.5). Above all, as W.G. Warren says, know your data.

12 LITERATURE CITED

Batschelet, E. 1981. Circular statistics in biology. Mathematics in Biology, R. Sibson & J. Cohen (Ser. Ed.). London: Academic Press. xvi+371 pp.

Bell, K. N. I. 1994. Life cycle, early life history, fisheries and recruitment dynamics of diadromous gobies of Dominica, W.I., emphasising *Sicydium punctatum* Perugia. Ph.D. thesis, Memorial Univ. of Nfld. St. John's, Nfld., Canada A1B 3X9. xviii + 275 pp.

Bell, K. N. I. 1997. Complex recruitment dynamics with Doppler-like effects caused by shifts and cycles in age-at-recruitment. Can. J. Fish. Aquat. Sci. 54: 1668-1681.

Bell, K. N. I. 1999. An Overview of Goby-Fry Fisheries. NAGA - the ICLARM quarterly 22: 30-36.

Bell, K. N. I. 2004. Introduction to circular variables & periodic regression in biology. (1st edition). Razorbill Press. 54 pp (electronic book in PDF format). ISBN 0-9736209-0-0.

Bell, K. N. I. 2007. Opportunities in stream drift: methods, goby larval types, temporal cycles, in situ mortality estimation, and conservation implications. *In* N. L. Evenhuis & J. M. Fitzsimons (Eds), Proceedings of the symposium on the biology of Hawaiian streams and estuaries. Bishop Museum Bulletin in Cultural and Environmental Studies **3** (pp. 35-61). Hilo, Hawai'i, 26–27 April 2005. Bishop Museum (Honolulu).

Bell, K. N. I., Cowley, P. D. and Whitfield, A. K. 2001. Seasonality in frequency of marine access to an intermittently open estuary: implications for recruitment strategies. Estuar. Cstl. Shelf Sci. 52: 327-337.

Bell, K. N. I., Pepin, P. and Brown, J. A. 1995. Seasonal, inverse cycling of length- and age-at-recruitment in the diadromous gobies *Sicydium punctatum* and *Sicydium antillarum* (Pisces) in Dominica, West Indies. Can. J. Fish. Aquat. Sci. 52: 1535-1545.

Bliss, C. I. 1958. Periodic regression in biology and climatology. Bull. Conn. Agric. Exp. Station, New Haven 615: 1-55.

Bliss, C. I. 1970. Statistics in Biology, Vol. 2. New York: McGraw-Hill. 639 pp.

CBE Style Manual Committee 1983. CBE Style Manual: a guide for authors, editors and publishers in the biological sciences. 5th ed.rev. and expanded. Chicago: Council of Biology Editors, Inc. xx + 324. ISBN 0-914340-04-2.

Cook, R. D. and Weisberg, S. 1982. Residuals and influence in regression. Monographs on statistics and applied probability, D. R. Cox & D. V. Hinkley (Ser. Ed.). London: Chapman and Hall. viii+30. ISBN 0-412-24280-X.

Cowley, P. D., Whitfield, A. K. and Bell, K. N. I. 2001. The surf zone ichthyoplankton adjacent to an intermittently open estuary, with evidence of recruitment during overwash events. Estuar. Cstl. Shelf Sci. 52: 339-348.

Fisher, N. I. 1993. Statistical analysis of circular data. Cambridge: Cambridge University Press. xvii + 277 pp.

Höfer, T., Przyrembel, H. and Verleger, S. 2004. New evidence for the Theory of the Stork. Paediatric and Perinatal Epidemiology 18: 88-92.

Hurlbert, S. H. 1984. Pseudoreplication and the design of ecological field experiments. Ecol. Monogr. 54: 187-211.

Landers, J. and Mouzas, A. 1988. Burial Seasonality and Causes of Death in London 1670-1819. Population Studies 42: 59-83.

Ricker, W. E. 1975. Computation and interpretation of biological statistics of fish populations. Bull. Fish. Res. Board Can. 191: xviii + 382 p.

Ricker, W. E. 1984. Computation and uses of central trend lines. Can. J. Fish. Aquat. Sci. 62: 1897-1905.

Snow, C. P. 1962. Science and government (from the Godkin lectures at Harvard University, 1960). New York: New American Library / Mentor (by arrangement with Harvard University Press). vii+128 pp.

Warren, W. G. 1986. On the presentation of statistical analysis: reason or ritual. Can. J. For. Res. 16: 1185-1191.

Wilimovsky, N. J. 1990. Misuses of the term "Julian Day". Trans. Amer. Fish. Soc. 119: 162.

Wilkinson, L. 1987. SYSTAT: The system for statistics. Evanston, IL: Systat Inc. ~400 pp (numbered by section).

Zar, J. H. 1984. Biostatistical analysis. (Second edition). New Jersey: Prentice-Hall. xiv+718 pp.

13 NOTATION

The notation used is sometimes new, so this section has a main-level heading for ready accessibility. Clear notation helps keep track of the formats and conversions, and thus helps to avoid errors. See also the section Special marks (Sec. 14.3), and Short list: main symbols and special notation (p. xv).

Where brackets aren't required for clarity, I may omit them; e.g. sinα or sin$x$ means sin(α) or sin(x). Because many variable names are acronymic, multiplication is often explicitly indicated by "*" (except where one part is numeric like "2DOY" which means 2*DOY) or use of brackets to avoid ambiguity. Subscripts that are obvious abbreviations of common words may be used without definition, and refer the subscripted variable to that word; e.g. SS_{resid} means the SS pertaining to residuals, i.e. residual SS.

In brief: the symbols used uniquely here are: the prime or single quote (') indicating a provisional or uncorrected value; the grave accent (` ) indicating the *proper* conversion or transform; and the double prime or double quote ( " or " ) indicating the ($x$",$y$") or ($X$",$Y$") conceptual coordinate space implied by any circular $X$ and derived as (sin`X,cos`X).

13.1 NOTATION: THE PROVISIONAL UNCORRECTED RESULT (')

When a calculation delivers a result that is ambiguous and requires correction to make it true, the uncorrected result is denoted by a following prime or single quote ('). For example, ∂' denotes an uncorrected lag, and α' an uncorrected angle (Sec. 6.4.4, 4.3, Eq. 6-9).

13.2 NOTATION: THE *PROPER* ANGULAR TRANSFORMATION (` or R`)

It is a simple mistake to try to take sin(60°) but using a function that expects radians, getting a wrong answer (e.g. -0.3) instead of 0.86. Notation helps keep track of what you've done, or intend to do.

We need conversions because computer programs usually reply to "sin(x)" assuming that the angle is given in radians, i.e. assuming $k=2\pi$ because there are 2π radians in a full cycle. (Programmers have avoided more complexity, assuming that operators can figure it out.) That means we have to convert the angular units of the observation (e.g. hour of day, day of lunar month, day of year) to radians. The conversion method is given in Sec. 3.5, but we do need to keep track of what we're doing.

No matter what units (k) a cycle is measured in, i.e. whether a day is measured as x of $k=1440$ minutes or as x of $k=24$ hours, the sine and cosine of any particular time are fixed. The notion of the *proper* sine and cosine conveys this concept. E.g.: taking midnight as the arbitrary zero, the sine of 3 p.m. (15 hours or 900 minutes into a 24 hour or 1440 minute day) is -0.707. The sine and cosine pair integrates both the cycle length (or scale) k and observation x, and conversion amongst units does not change the sine (etc.) of the observation.

It's helpful to think of an angle as its k and x together as (k,x), so that for example, still using 3 p.m. as above, $\sin(24,15) = \sin(1440,900) = \sin(1,0.625)$, and in radians $= \sin(2\pi,2\pi*0.625) = -0.707$. If that seems miraculous, it's only because we're accustomed to not writing k and we forget that we have assumed k in the sine (etc.) tables we use; but it arises directly from the geometric definition of sines, cosines, etc.; see Sec. 3.2.3, Figure 3-1, Figure 3-3. $\text{Sin}(k,x)$ is available as a notation or form because there's no similar-looking use; $\sin(k,x)$ is equivalent to our notation $\sin\grave{}x$.

Clear notation helps reduce mistakes and trace errors. The notation we use is the grave symbol " $\grave{}$ " (or $\mathbf{R}\grave{}$, $\mathbf{R}$ followed by the 'grave' symbol) preceding the units x, to imply the proper conversion to standard angular units (or particularly to radians) that allow properly taking the sine, cosine, etc. Cycle length of such standard angular measures can be indicated as $k_{standard}$. Thus:

> $\grave{}$ or $\mathbf{R}\grave{} = (k_{standard}/k)$ or $(2\pi/k)$, e.g. $\grave{}x = (k_{standard}/k)*x$. Notation $\grave{}x$ means
> "x properly converted to the standard angular measure used to obtain sine, cosine, etc." Notation $\mathbf{R}\grave{}x$ means "x properly converted to radians". For example, $\mathbf{sin}\grave{}x$ means $\mathbf{sin}[(k_{standard}/k)*x]$, the *proper* sine of x. The notation
> lets us write $\sin\grave{}$TOD (etc.), and read it as "[proper] sine of time of day" etc.; and succinctly identify the variable without letting units of measure encumber communication of (or implementation of) concepts.

When writing α in standard angular units from which we can take sine etc. directly, we may omit the grave notation; likewise anywhere the meaning is clear.

13.3 NOTATION: THE $(X'',Y'')=(\text{SIN}\grave{}X,\text{COS}\grave{}X)$ COORDINATE SYSTEM

A double-prime or double-quote notation explicitly identifies a (conventional) coordinate system $(X'',Y'')=(\sin\grave{}X,\cos\grave{}X)$; i.e. both X'' and Y'' are derived from X. The variables marked with a double prime are (here) always independent variables.

Although we could use un-ornamented (x,y) as Batschelet (1981) did (e.g. his p. 236), that is potentially confusing. The (x'',y'') notation accommodates cycles while retaining the un-ornamented y or Y for the dependent variable (avoiding the need to use an x,y,z system with z the independent variable). The notation is optional however, and can be dispensed with where the meaning is otherwise clear.

When we have a situation like observed Y vs. a circular X, such as growth vs. DOY, we have what is often called a "cylindrical" situation, because the X, being circular, is essentially two-dimensional. It is possible to treat X as cyclic, and plot, say, Y vs. degrees or DOY (e.g. Figure 6-1). Fundamentally, though, the circular variable X implies a (sin`$X$, cos`X) coordinate system (as in Figure 6-3). Indeed, analysis requires the proxy variables (sin`$X$, cos`X), which—for convenience and clarity—we may rename (X'',Y''), with points (x'',y''), where the double quotes indicate that they are both proxy variables of the original circular X variable.

X'' and Y'' (X or Y double prime or double quote) notation can include uses where $X''=L*$sin`$X$, where $L$ is vector magnitude or variable, e.g. for coordinates of vectors. Notations can become awkward, for example: we can, for a polar plot (Sec. 6.5.2), write $X''=Y*$sin`X, but it means $X''=Y*X''$, which would be fine as a line of code but might puzzle some readers; so $X_{polar}=Y*x''$ might be preferable.

Why the (x'',y'') notation? Yes, I could have chosen an X,Y,Z system, but: • X,Y,Z still would have confused the transformed X with the linear X; • I wanted to retain Y as a dependent variable; • it's useful that (X'',Y'') clearly signify a pair and thus help reinforce their origin from a single X. The notation is also extensible to multiple X variables, as $X1'',Y1''$, and $X2'',Y2''$, etc.

13.4 NOTATION & CALCULATION OF HARMONICS

Twang a guitar string and you get a certain note (its fundamental or first harmonic) that depends on its weight and tension. But block the string halfway with your thumb while you twang it with a finger, and you get a twice-higher frequency, the second harmonic. (You can also twang 1/3, 1/4 etc. of the string, to get 3rd, 4th, etc. harmonics.) Just as the string vibrates sinusoidally (the position of a mark on the vibrating string will trace a sinusoidal curve when plotted over time), physical and biological variables also vary sinusoidally over seasons, times of day, etc. Harmonics in music and nature are thus analogous.

A single complete cycle of variation over its fundamental time period (k) is the first harmonic. Harmonics can be thought of as other integer numbers (n) of complete cycles in the same time period, or as cycles with periods k/n. From the index x you can generate simultaneous multiple sets of (sin,cos) for whichever harmonics are of interest.

Notation and calculation: harmonics can be flagged by the number of harmonic cycles between the trigonometric function's name and the variable, e.g., working with radians (R), sinR`2DOY, indicating the sine is to be taken of the second harmonic of time of year, being a 182.5 d cycle. The "2" in "sin($2x$)" or "sinR`2DOY" means that the angle (x or DOY) is doubled, or the factor R` is doubled $(2\pi/(k/2) = 2*2\pi/k)$, which is equivalent to our intention that k is divided by 2. Calculating a harmonic can thus be quickly done by multiplying the index x by the order of the harmonic, e.g. if $x = 70°$, then $\sin(70°)=0.94$, and the sine of that x on the second harmonic is $\sin(2*70°)=0.64$.   Notation like "sinR`2DOY" is therefore not a mere indicator, but a mathematically valid expression for harmonics.

Remember that trigonometric transformations are not like arithmetical operators, so sinR`2DOY ≠ 2sinR`DOY. The R` or ` not only signifies the proper transform of x, but it functions like a pair of brackets enclosing the following term.

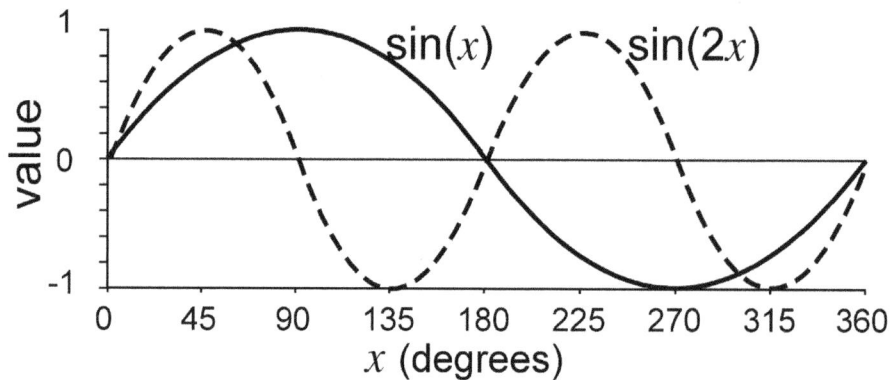

Figure 13-1. Harmonics. The nth harmonic undergoes n complete cycles in the space or time that the first harmonic undergoes just one. Sin(x) or cos(x) on the nth harmonic equals $\sin(nx)$ or $\cos(nx)$.   That is somewhat counterintuitive, but works because $\sin(2x)$ describes a complete cycle in $k/2$ units of the index x (for this example, $180°=360°/2$).

The harmonics you construct in your spreadsheet will consist of paired columns (sin,cos) of proxy Xs for each harmonic, just as for the first harmonic (see examples in Sec. 9).

14 GLOSSARY: ABBREVIATIONS, SYMBOLS, MARKS

This list is intended to explain terms especially important in this manual. Terms are grouped in themes (General, Standard symbols, Special marks), which I hope will help more often than it hinders. Notation is dealt with in Sec. 13.

> Notes: [1] Some names are acronymic; multiplication may therefore be explicitly indicated by "*" (except where one part is numeric like "2DOY" which means 2*DOY) or use of brackets where it is otherwise ambiguous. [2] subscripts that are obvious abbreviations of common words may be used without definition, and refer the subscripted variable to that word; e.g. SS_{resid} means the SS pertaining to residuals, i.e. residual SS.

14.1 GENERAL

* (asterisk): multiplication operator.

^ or ** (caret or double asterisk): alone, exponentiation operator. See also $\hat{y}$.

∂, see delta, phase.

A, see amplitude.

AAR, Age-at-Recruitment: see MLD.

acrophase: see phase angle.

arcsin, arccos, arctan: inverses of the trigonometric functions sine, cosine, and tangent. (Also written $\sin^{-1}$, $\cos^{-1}$, $\tan^{-1}$). Sec. 3, 4.3.

arctan method: finds true angle α by arctan(sinα/cosα)+QC. Sec. 4.3.

amplitude (A): in units of Y, absolute distance from the mesor to maximum or minimum of a periodic function. Calculated by Pythagoras' theorem $A = \sqrt{(B_{sin}^2 + B_{cos}^2)}$ from regression coefficients for the cycle. Measure of relative importance (not significance) amongst cycles in a regression. Range=2A. Sec. 6.

average: see mean (arithmetic).

azi: azimuthal system. Sec. 3.2.1.

c: Bliss's (1958) notation for $2\pi/k$.

category [variable]: a qualitative identifier that groups observations for comparison or separate analysis.

circular average: see mean angle; cf. resultant angle.

circular coordinates: (x",y") express an angle x on a unit circle, where $x"_i = \sin`x_i$ and $y"_i = \cos`x_i$. This notation conveys 2 dimensionality of cycles and reserves plain Y or y for the dependent variable. (It allows an x",y",y (x double-prime, y double-prime, y) system instead of x,y,z). Sec. 13.3.

circular variable: variable like direction or time of day that recurs as, or implies, a cycle and is often an index (X) against which periodic phenomena can be analysed; See format, periodic variable.

common format: see formats.

cosine (cos): in a right triangle, a trigonometric function of acute angle such that cos(α) = adjacent_side /hypotenuse. Inverse: see arcsin etc. See sine, tan, trigonometric function, Sec. 3, 4.3.

cosinor analysis: term (avoided in this book) for the "cosine form" (Sec. 6.4.2) of periodic regression.

cycle: (*see also* period) a repeating sequence from 0 to k. 2nd, 3rd, etc. harmonics are subsidiary cycles with 1/2, 1/3, etc. of the primary period. Values x in cycles can be expressed as angles (e.g. degrees, radians ...) or non-standard angles (e.g. year as days, months, ... seconds), or, fundamentally, in trigonometric format (sin`x,cos`x). *See* format.

cyclic format: non-standard angular representation of a circular variable, using a scale from 0 to k, e.g. DOY (0 to 365). *See* format; Sec. 3.4.

day of year: *see* DOY.

delta, ∂: lag from t_0 to the cycle's peak; phase angle in periodic regression. *See* phase, & Sec. 6, 6.3, 6.4.4.

degrees of freedom (DF, d.f., *df*, etc.): a system for accounting for the number of data and the number of parameters estimated from them (Sec. 11.3.7), and used in finding the appropriate reference value of a test statistic for inference. In subscript, *df1* and *df2* reflect the regression or groups (numerator, $=m$) and residual (denominator, $=n-m-1$) DF.

dependent variable: the variable, usually written Y, presumed affected by an independent variable X.

DF, d.f., *df,* etc: *see* degrees of freedom.

DOY: Day-of-Year, 0-365. Decimal portion indicates part of day (i.e: "DOY = 0.5" means 12 noon on Jan. 01; 365.0 is the start of the next year). "Julian Day" does not mean day-of-year; its use as such is an illiteracy. Sec. 5.3.1, 3.4; DOY.

effect: association, not necessarily causal despite common meaning of "effect", of X or Xs and Y.

fitted values, ¥, $\hat{y}$ (read y-hat) or y_{fitted}: the expected y_i values that can be calculated from the regression equation at each *observed* value x_i. (*cf*. predicted values).

format, circular: a circular variable can be represented in 2-dimensional *trigonometric* format (sin,cos); or in 1-dimensional or *angular* format as (e.g.) degrees (0 to 360); or *cyclic* formats (non-standard angles, e.g. DOY (0 to 365)); *common* or everyday formats, often not mathematically proper (lack zero, e.g., Jan 0 does not exist); *linear* format expresses continuous sense (time, rotation).

harmonic: the nth harmonic is a repeating sequence having period length $1/n$ of a (larger) cycle. Sec. 13.4.

hhmm: 24-hour format (often called European or military time) for time of day; 2:30 pm in this format is written 1430h, meaning 14 hours 30 minutes after the arbitrary zero.

hypothesis, H: an assertion to be tested. H_0: *See* null. Sec. 11.3.6, 11.3.5, 11.1.7.

independent or predictor variable: the X variable, presumed to affect the dependent variable Y. *See* Sec. 11.1.7.

index X, or index variable: an independent variable (usually time) to which an observation is indexed, e.g: times x or depths x etc., at which y was measured. Index may be circular (time of day) or linear (e.g. time since a mark or event, as with calendar date and time). Sec. 3.1.1.

Julian Day: a continuous time system in astronomy; it does not mean day-of-year and such usage is usually wrong. *See* DOY (Day-of-Year) under Sec. 5.3.1.

k = length (period) of cycle in whatever units are used. Standard cycles have $k_{standard}=2\pi$ (radians), or 360 (degrees). k can reflect common units as observed (e.g. 365 days, 24 h, etc.). Convert x to scale B from A by ratio k_B/k_A. Sec. 3.2.

L: length of a vector extending from $(0,0)$ to (x,y); $L = \sqrt{(x^2+y^2)}$. Sec. 4.1.

lag: the difference in the index variable between one thing and another; e.g. the phase angle and the arbitrary zero of the cycle (phase – t_0).

linear: interval or ratio data type; monotonic format of a circular (date or angle) variable.

MA: *see* mean angle.

mean: a measure of central tendency, with 3 kinds that are of interest here:
(periodic): *see* mesor.
(arithmetic): average, $(\sum x)/n$, the nth part of the sum of n data.
(geometric): $(\prod x)^{\wedge}(1/x)$, the nth root of the product of n data.

mean angle (MA) or circular average: average direction, the angle defined by $(\sum \sin\alpha, \sum \cos\alpha)$ of n angles. Does not equate to the arithmetic mean. Can equal resultant angle (RA, of MV or RV) only if RA is based on angles or unit vectors. Sec. 4.1.

mean vector, MV: the average direction and magnitude of n vectors. The terminal coordinates are those of the resultant vector $(\sum L\sin\alpha, \sum L\cos\alpha)$ divided by n, thus it has the same angle as the RV. Its length is RVL/n, or calculated from MV coordinates using Pythagoras' theorem. Sec. 4.1.

mesor, M (or periodic mean): expected mean value of Y over (or taking into account) a cycle; generally superior to the mean of Y. In regression it functions like an intercept. Originally an acronym for Midline Estimating Statistic Of Rhythm. Sec. 6, 6.4.4.

MLD, or PLD, OLD (Marine, Pelagic, or Oceanic Larval Duration), or (*see*) AAR: life-history term. Usually, number of days in ocean from hatch to recruitment†, but the concept can extend to the duration of any stage. †Recruitment means entry of a fish into a new category, e.g. the 'fishable' category, or return to fresh waters (for juvenile-return anadromous (amphidromous) species).

MV: *see* mean vector.

MVL: length of mean vector.

nDOY: days since 0000h of Jan 01 or an arbitrary year. *See* DOY. sin`DOY=sin`nDOY, etc.

normal: a symmetrical distribution, the Gaussian or (popularly) "bell curve"; data conforming to same. The standard normal has $\mu = 0$ and $\sigma^2 = 1$. The sum of multiple random processes tends to be normally distributed (Sec. 11.1.4).

null: (hypothesis, H_0) of no difference or effect; (population) where effect being tested does not exist, used to test sample data . *See* hypothesis; Sec. 11.3.6, 11.3.5, 11.1.7.

orthogonal: a quality of an analysis design. Sec. 6.4.6.

peak: in a cycle or periodic function, the maximum of Y. *Cf.* phase angle.

period (*see also cycle*): length over which a pattern repeats. *See k*.

periodic mean: *see* mesor.

periodic regression: regression of a linear Y against a circular X. A geometric analogy is a cylinder with Y on the height of the cylinder, and X on circumference. Sec. 6.

periodic variable: one showing a pattern of rising and falling with regularity, often according to some circular variable like time of year, etc. The sin & cos transforms of a circular variable are periodic. *See* circular variable.

phase [angle] (also: acrophase [angle], peak location (esp. when referring to phase in common format), delta or ∂ (in context)): in a cycle, the value of X, or position, where the contribution of X to Y is greatest; calculated from the cycle's pair (sin and cos) of coefficients in a periodic regression. Sec. 6, 6.4.4.

pol: polar angle system. Sec. 3.2.1.

predicted values: as for fitted values, but at X values not observed, or beyond the range of observations.

proper (sine, cosine, etc.): the sine (etc.) of the appropriately-transformed variable, i.e. to radians, degrees, etc. as required for those functions. "Proper" is indicated by " ` ", e.g. sin`x, read as proper sine of x. *See* R`, `, Sec. 13, 14.3.

pseudo-random: a numerical process of generating a random-like number. The process begins with a 'seed' number, often a digit from the current time; many arbitrary calculations follow, like 'square x, then take the 4th digit after the decimal', etc., to produce a pseudo-random result difficult to predict from the 'seed'. *See* random.

QC: *see* Quadrant correction.

Quadrant: the first, second, third, or fourth quarter of a cycle. *See* Figure 3-2, Figure 3-3.

Quadrant correction (QC): a correction used to obtain true angle α or ∂ from coefficients (B_1, B_2) or coordinates (sin,cos); ∂' or $\alpha' = \arctan(x''/y'')$, and true $\alpha = \alpha' + QC$. Sec. 4.3, 10.1.3, 10.2.1.

RA: *see* resultant angle.

radians (rads): angular unit; 1 rad = $1/2\pi$ cycles, and $k = 2\pi$, i.e. 2π rads = 1 cycle. Most computer routines assume radians. Angular measures can inter-converted by the ratios of their ks, e.g., from α_k in cycle of period k units, $\alpha_{rads} = \alpha_k * (2\pi/k)$, and $\alpha_{degrees} = \alpha_k * (360/k)$. Sec. 3.5.

random: unpredictable, arising by chance (*see* pseudo-random); ~number: one having no clear relation to anything; essentially unpredictable. Coin tosses and dice throws are sufficiently unpredictable to be called random. Sec. 11.1.4. ~sample: chosen by a random method, in order to disassociate sample choice from properties of interest. Sec. 11.3.3, 1.2.3.

range: in a periodic function, the range of the contribution (maximum-minimum, through the cycle) to Y by any cyclic X; twice the amplitude A.

recruitment: generally, a transition from one category into another, or the numbers so transitioning, e.g. recruitment into the army. In fisheries/ecology, it is the process or time of becoming vulnerable to a [defined] fishery, or entering a new habitat or life history stage.

regression: the statistical process of relating a Y to an X, or the result thereof (Sec. 11.1.7), or the equation so obtained.

residuals, ε_i, e_i, or errors: $(y_i - \hat{y}_i)$, $y_{observed}$ minus y_{fitted}, the component of variation of Y that is unaccounted for by a regression or other model (Sec. 11.1.7). Better models have residuals that: show less pattern (vs. any X in the model); are smaller ($\sum$residuals^2 is a standard criterion of fit); are normally distributed. Sec. 11.1.7.

resultant angle, RA: angle of either resultant or mean vector. RA equals mean angle only if the vector is unitary.

resultant vector, RV: the net result of n vectors i, and having terminus $(\sum L_i \sin\alpha_i, \sum L_i \cos\alpha_i)$, length RVL = $((\sum L_i \sin\alpha_i)^2 + (\sum L_i \cos\alpha_i)^2)^{0.5}$, and RA (calculated as in Sec. 4.3). *See* vector.

RV: *see* resultant vector.

RVL: length of resultant vector.

sample: a set of individuals taken from a population. Sec. 11.1.1, 11.1.7, 11.3.2, 11.3.3.

significance: a value expressed by (*see*) Sec. 11.2.3, 11.1.7.

[statistically] significant: in reference to a test statistic calculated from data, a lower-than-critical-level probability of finding an equivalent result under the null hypothesis H_0, thus signifying inconsistency with the null and justifying its rejection. Signifying (via statistical test) some effect beyond chance. Sec. 11.2.3, 11.3.1, 11.3.5.

sine (sin): in a right triangle, a trigonometric fn. of acute angle such that $\sin(\alpha)$ = opposite/hypotenuse. *See* proper, cos, tan, trigonometric function, Sec. 3, 4.3.

standard angular units: those (degrees, radians, or grads) accepted by tables, calculators, or computers giving transforms sine, cosine, tangent, etc.

tangent (tan): trigonometric function defined as ratio opposite/adjacent (sides in right triangle); used to find angle from coordinates (Sec. 4.3). *See also* trig. function, cos, sin, Sec. 3.

trajectory: the path created by joining n vectors in order. (Sec. 4.1.) *Cf.* resultant vector, which is order-independent.

trigonometric function (of angle): one like (*see*) sine, cosine, tangent. *See also*: proper, & Sec. 3.1.2.

t_0: nominal zero time; the zero of a circular variable is arbitrary. (Sec. 1.1.)

UTC, GMT: Universal Time Coordinate or Greenwich Mean Time. The time at the zero longitude meridian (on which Greenwich lies). Sec. 5.3.2.

unit vector: one having L = 1.0.

unitary (Resultant or Mean) vector: one calculated from unit vectors.

vector: (geometry) a directed line with length L and angle α from which $(L^*\sin\alpha, L^*\cos\alpha) = (x',y'')$, its circular coordinates (Sec. 13.3); $L=\sqrt{(x'^2+y''^2)}$ and $\alpha= \arctan(x''/y'')$+QC (Sec. 4.3). Translate to arbitrary origin (x_A,y_A) by adding a vector (x_A,y_A). *See* unit, mean, resultant v.; Sec. 4.1.

14.2 STANDARD SYMBOLS

Letters are convenient, but due to their limited supply each finds many different uses pointed to by context.

α, alpha: (statistics) a critical level of probability, below which a result is deemed significant, i.e. signifying a non-chance effect. (Trigonometry) α is an angle. Sec. 11.1.

Δ, read "delta": a change or difference in (two states, conditions, versions, etc. of) the quantity following.

$\sum$ (Greek sigma, uppercase): summation operator. Unless specified, the entire range of the subscript is to be summed, e.g. $i=1$ to n.

σ (Greek sigma, lowercase): a measure of dispersion, the population standard deviation. σ^2 = variance. *Cf. s.*

μ (Greek mu, lowercase, italic): the population mean of x, or of y if so subscripted. *Cf.* $\bar{x}$, $\bar{y}$.

B_0, B_1, B_2 ... B_n: regression coefficients. B_0 is the intercept, (mesor in periodic regression [PR]); B_1 associates with X1 (in PR usually $\sin`X$); $B_2$ associates with X2 (in PR usually $\cos`X$). Cycles are thus represented by *pairs* of Bs.

ε_i, e_i, or errors: *see* residuals.

DF, d.f., *df*, etc: *see* degrees of freedom.

F: Fisher's F, a test statistic; DF are usually declared with F. Sec. 11.2.3, 11.3.6, 11.3.7.

H, hypothesis: *see* null.

MS: a mean square, SS/DF, (total, regression, or residual MS identified by subscript or context). *See* SS.

m: number of predictors, independent variables, for which coefficients are to be estimated in a model. *Cf.* DF.

n: total number of data. *Cf.* DF.

p, $p[event]$: probability, [of *event*]; $p[F]$ is p[an equally or more extreme F occurring under the null hypothesis]. Low p implies high significance; p does not have DF, although its source statistics usually do. Sec. 11.2.3, 11.1.7.

P: peak (location of). Sec. 6.3, 6.4

r: correlation coefficient. See R^2.

R^2: coefficient of determination or squared multiple R. 100*R^2 is usually equated to percent of variation explained. Sec. 11.2.3.

s, and s^2: sample standard deviation and variance. *Cf.* σ.

SS: sum(s) of squares (total, regression, or residual SS identified by subscript or context). *See* MS. Sec. 11.1.7, (DF) 11.3.7.

t: 1. Student's *t*, a test statistic; DF are usually declared with *t*. 2. Time (*see* t_0).

variance: see s^2, σ^2.

X and *x:* independent variables (typically uppercase), or observations (lowercase). E.g., x_i is *i*th observation of *X*. Circular *X*s have formats including trigonometric: (sin`x,cos`x), or (*x*",*y*"). Variables can be numbered (*X1*, *X2*, etc.).

Y and *y:* dependent variables (uppercase) and observations (lowercase). E.g., y_i is the *i*th observed value of *Y*. Don't confuse *Y* or *y* (dependent variable) with (*see*) *Y*' or *y*". *See X*.

$\bar{x}$ (*x*-bar): mean of *x* (sample). *Cf.* μ.

$\bar{y}$ (*y*-bar): mean of *y* (sample). *Cf.* μ_y.

$\hat{y}$ (y-hat) or y_{fitted}: *see* fitted value.

14.3 SPECIAL MARKS

Marks used uniquely in this book. *See also* Sec. 13, Notation.

`` ` `` : grave accent (*see* grave).

' : *see* prime .

" : *see* double-prime .

`X: indicates circular variable *X* transformed to standard angle (deg, rad, etc.). *See* R`X, grave.

double-prime, ", or double quote, as in (*x*",*y*"), indicates coordinate space (sin`x,cos`x) implied by an angle or

time *x*. *See* circular coordinates, *X*", Sec. 13.3.

grave (`` ` ``) accent (pronounced '*grahv*'): operator of proper transformation to unspecified standard angular units, allowing sine, cosine, etc. to be properly applied. Often preceded by trigonometric operator (e.g. sin`DOY or sinR`DOY). Form R` specifically denotes use of radians, equivalent to Bliss's (1970) "*c*". R` and `` ` `` can be read, as in sin`DOY, as "proper" sine (etc.).
R`=2π/k_x, but grave (`` ` ``) alone can have numerator for unspecified standard *k* (deg, rad, grad...). E.g., sinR`0430h (24h time) means, using radians, sin(4.5*2π/24). *See* R`, R`X, `X, & Sec. 13.1.

prime, ', or straight single quote: following an estimated value it indicates a provisional, uncorrected value, e.g. ∂' or α' not yet corrected for quadrant (Sec. 4.3).

R`: indicates radians-transformation; R`=2π/*k*. *See* grave.

R`X: indicates radians-transformed circular variable *X*. *See* grave.

R`DOY [or `DOY]: DOY (day of year, *k*=365) transformed to radians [or any standard units]. *See* grave.

R`TOD [or `TOD]: TOD (time of day, with *k* = 24 h or 1 cycle) transformed to radians [or any standard units]; R`TOD = TOD*2π/*k*. *See* grave.

$_{[period]}$`[variable]: optional subscript preceding grave mark `` ` `` to show period (*k*) of *X* index variable, e.g. sin_{24}`15.5 means proper sine of 15.5 units in cycle of *k*=24. *See* grave (`` ` ``).

X", *x*", *Y*", *y*": double-prime mark " denotes circular coordinates (*x*",*y*") of circular index (*X*) variable. $x"_i$ = sin`$x_i$, $y"_i$ = cos`x_i. *See:* " (mark), double-prime, *X*.

¥, y_{fitted} or $\hat{y}$: *see* fitted values.

15 INDEX

Bold page numbers are key entries. *See also* Glossary (it is not indexed).

16 DAY-OF-YEAR TABLE (DOY 0-365)

Table 16-I. Year begins at 0, not 1. (Although some have used DOY as 1 to 365, that's mathematically nonsensical because 1200h Jan 01 is DOY 0.5, not 1.5, likewise 1200h Dec 31 is DOY 364.5, not 365.5.) For leap year Feb 29 is DOY 59, and add 1 to each DOY from Mar 01; nevertheless, the biological meaning of DOY is Earth's position in the solar orbit, and leap year corrections are too small, $\leq 0.25/365$, to matter for most purposes.

To convert hhmm to Decimal part of Day (DD): DD=hh/24 + mm/1440. E.g.: 16.28 {hh.mm} is 16/24 + 28/1440 = 0.686 {proper DD}. Thus, 1628hrs on Nov. 15 = DOY.dd 318.686. See Sec. 10 for automation in spreadsheets.

d	*DOY*																
		10	40	24	82	4	123	15	165	27	207	6	248	18	290	29	332
JAN		11	41	25	83	5	124	16	166	28	208	7	249	19	291	30	333
1	0	12	42	26	84	6	125	17	167	29	209	8	250	20	292	**DEC**	
2	1	13	43	27	85	7	126	18	168	30	210	9	251	21	293	1	334
3	2	14	44	28	86	8	127	19	169	31	211	10	252	22	294	2	335
4	3	15	45	29	87	9	128	20	170	**AUG**		11	253	23	295	3	336
5	4	16	46	30	88	10	129	21	171	1	212	12	254	24	296	4	337
6	5	17	47	31	89	11	130	22	172	2	213	13	255	25	297	5	338
7	6	18	48	**APR**		12	131	23	173	3	214	14	256	26	298	6	339
8	7	19	49	1	90	13	132	24	174	4	215	15	257	27	299	7	340
9	8	20	50	2	91	14	133	25	175	5	216	16	258	28	300	8	341
10	9	21	51	3	92	15	134	26	176	6	217	17	259	29	301	9	342
11	10	22	52	4	93	16	135	27	177	7	218	18	260	30	302	10	343
12	11	23	53	5	94	17	136	28	178	8	219	19	261	31	303	11	344
13	12	24	54	6	95	18	137	29	179	9	220	20	262	**NOV**		12	345
14	13	25	55	7	96	19	138	30	180	10	221	21	263	1	304	13	346
15	14	26	56	8	97	20	139	**JULY**		11	222	22	264	2	305	14	347
16	15	27	57	9	98	21	140	1	181	12	223	23	265	3	306	15	348
17	16	28	58	10	99	22	141	2	182	13	224	24	266	4	307	16	349
18	17	**MAR**		11	100	23	142	3	183	14	225	25	267	5	308	17	350
19	18	1	59	12	101	24	143	4	184	15	226	26	268	6	309	18	351
20	19	2	60	13	102	25	144	5	185	16	227	27	269	7	310	19	352
21	20	3	61	14	103	26	145	6	186	17	228	28	270	8	311	20	353
22	21	4	62	15	104	27	146	7	187	18	229	29	271	9	312	21	354
23	22	5	63	16	105	28	147	8	188	19	230	30	272	10	313	22	355
24	23	6	64	17	106	29	148	9	189	20	231	**OCT**		11	314	23	356
25	24	7	65	18	107	30	149	10	190	21	232	1	273	12	315	24	357
26	25	8	66	19	108	31	150	11	191	22	233	2	274	13	316	25	358
27	26	9	67	20	109	**JUNE**		12	192	23	234	3	275	14	317	26	359
28	27	10	68	21	110	1	151	13	193	24	235	4	276	15	318	27	360
29	28	11	69	22	111	2	152	14	194	25	236	5	277	16	319	28	361
30	29	12	70	23	112	3	153	15	195	26	237	6	278	17	320	29	362
31	30	13	71	24	113	4	154	16	196	27	238	7	279	18	321	30	363
FEB		14	72	25	114	5	155	17	197	28	239	8	280	19	322	31	364
1	31	15	73	26	115	6	156	18	198	29	240	9	281	20	323	End	of
2	32	16	74	27	116	7	157	19	199	30	241	10	282	21	324	year	=
3	33	17	75	28	117	8	158	20	200	31	242	11	283	22	325	365.0	=
4	34	18	76	29	118	9	159	21	201	**SEPT**		12	284	23	326	0.0 (start	
5	35	19	77	30	119	10	160	22	202	1	243	13	285	24	327	of next	
6	36	20	78	**MAY**		11	161	23	203	2	244	14	286	25	328	year)	
7	37	21	79	1	120	12	162	24	204	3	245	15	287	26	329		
8	38	22	80	2	121	13	163	25	205	4	246	16	288	27	330		
9	39	23	81	3	122	14	164	26	206	5	247	17	289	28	331		

www.ingramcontent.com/pod-product-compliance
Lightning Source LLC
Chambersburg PA
CBHW081504200326
41518CB00015B/2378